OBSERVABILITY AND OBSERVATION IN PHYSICAL SCIENCE

SYNTHESE LIBRARY

STUDIES IN EPISTEMOLOGY,

LOGIC, METHODOLOGY, AND PHILOSOPHY OF SCIENCE

VOLUME 209

PETER KOSSO
Department of Philosophy, Northwestern University, Evanston, U.S.A.

OBSERVABILITY AND OBSERVATION IN PHYSICAL SCIENCE

KLUWER ACADEMIC PUBLISHERS
DORDRECHT / BOSTON / LONDON

Library of Congress Cataloging in Publication Data

Kosso, Peter.
Observability and observation in physical science / Peter Kosso.
p. cm. -- (Synthese library ; v. 209)
Bibliography: p.
Includes index.
ISBN 0-7923-0389-X
1. Science--Observations. 2. Science--Methodology.
3. Experimental design. I. Title. II. Series.
Q175.K8649 1989
502.8--dc20 89-15494

ISBN 0-7923-0389-X

Published by Kluwer Academic Publishers,
P.O. Box 17, 3300 AA Dordrecht, The Netherlands.

Kluwer Academic Publishers incorporates
the publishing programmes of
D. Reidel, Martinus Nijhoff, Dr W. Junk and MTP Press.

Sold and distributed in the U.S.A. and Canada
by Kluwer Academic Publishers,
101 Philip Drive, Norwell, MA 02061, U.S.A.

In all other countries, sold and distributed
by Kluwer Academic Publishers Group,
P.O. Box 322, 3300 AH Dordrecht, The Netherlands.

printed on acid free paper

Printed in the Netherlands

CONTENTS

PREFACE

The concept of observability of entities in physical science is typically analyzed in terms of the nature and significance of a dichotomy between observables and unobservables. In this book, however, this categorization is resisted and observability is analyzed in a descriptive way in terms of the information which one can receive through interaction with objects in the world. The account of interaction and the transfer of information is done using applicable scientific theories. In this way the question of observability of scientific entities is put to science itself.

Several examples are presented which show how this interaction-information account of observability is done. It is demonstrated that observability has many dimensions which are in general orthogonal. The epistemic significance of these dimensions is explained.

This study is intended primarily as a method for understanding problems of observability rather than as a solution to those problems. The important issue of scientific realism and its relation to observability, however, demands attention. Hence, the implication of the interaction-information account for realism is drawn in terms of the epistemic significance of the dimensions of observability. This amounts to specifying what it is about good observations that make them objective evidence for scientific theories.

The method of analyzing observation and the epistemological conclusion will be of interest to anyone concerned with the philosophical claim that observation is theory-laden or that there is no clear distinction between observable and unobservable scientific entities. The issues discussed are relevant to the issue of scientific realism and to epistemology of science in general. The discussions are self-sufficient and there are no prerequisits, neither philosophical nor scientific. My intent is that it will be informative reading for professional philosophers of science as well as students at both the graduate and advanced undergraduate levels.

ACKNOWLEDGEMENTS

I have had help in completing this study of observation and observability. Philip Kitcher got me started and kept me going with his cheerful encouragement and his insight. Thanks.

Cindy has been with me from start to finish and that has made it all enjoyable. She has helped me with the pictures too. Thanks.

And thanks to the friendly people of the Montana State University Department of Physics who made me feel welcome and whose assistance gave me the time to finish the project. I am especially grateful to Linda Todd who did much of the hard work for this by helping me with the typing.

The research for this study was begun under the support of a Dissertation Fellowship given by the University of Minnesota Graduate School.

pk

INTRODUCTION

Observability in science is most often construed as a topic addressing the distinction between observable and unobservable scientific entities. Attitudes usually differ as to the epistemic significance of the dichotomy and the relevant criteria for determining how the distinction is to be drawn. But my concern will be neither with defending nor dissolving such a dichotomy, nor will it seek the proper application of the concept of being observable. Rather, the issue of observability will be more generally presented as an account of the flow and content of information from the world to human scientists. Under this interpretation, observability becomes primarily a descriptive part of the epistemology of science, and only subsequently prescriptive. One wants first to explain the details of just what is going on in the process of getting information from the world, and only then evaluate and classify. This priority will guide my approach to observability.

Why worry about observability? Perhaps it is an issue where resolution can be had simply by opening our eyes and unstopping our ears, and seeing or hearing what there is to see and hear. But it is not that simple. Observability cannot be decided on the basis of observation, since what is observable is not always observed. Planets of distant stars, for example, are, being relevantly similar to planets of our solar system, observable, though none have been observed. Furthermore, in many observations reports it is not clear what it is that is being observed, that is, the informational content of the observation. If we observe the table, can we in the same act claim to observe its constituent molecules? Clearly, there is some tracking of information to be done to understand observation and observability.

This tracking of information will be an important service for the issue of scientific realism. An important aspect of that debate concerns the use of observability characteristics as significant criteria in the warrant for belief in the existence of entities. The empiricist regards observability as epistemically very significant and licenses belief in existential claims as a function of the observability, appropriately defined, of the entities in question. The realist typically denies this epistemic significance of observability and ascribes belief by criteria which can be independent of observability. Clearly, this aspect of the realism issue has a great deal invested in the nature of observability and we are sure to be closer to a resolution of the debate by detailing the nature of observability.

Understanding observability is also important to evaluating the objectivity and rationality of the scientific business. Observation, as the informational link from the physical world, is the likely source of an objective foundation for scientific claims. It is common sense to claim that if one wants to know what is true of the world one ought to look first to the world, that is, to anchor one's claims to what is observable. Furthermore, look always to the basic observations to decide between competing theories, and thereby allow what is observable to count as the rational criteria for accepting theories. There is common sense in these claims, but with all the weight they place on the notions of observation and osbervability, there is also a common sense in wanting a careful understanding of those concepts.

The approach of this study of observation and observability will be largely empirical in that a variety of examples of observation and observability claims in physical science will be used as evidence to describe the informational nature of those activities. The conclusion of this study will be largely methodological in that what will be emphasized is a method of studying observability more than a univocal solution to the question of observability. The motivation for this emphasis comes form the examples themselves in that the variety of examples presents a variety of important aspects of observability. More can be learned from an attitude of investigation, asking after the aspects of information about the world for each case, than from an attitude of trying to fit all cases into a single shape. It will not do to draw generalizations about observability from just a small number of paradigm cases since this presupposes a unity which may not exist. It is only by understanding observability in many particular cases that one is licensed to speak of observability in general.

The initial motivation for studying a variety of examples as a way to understand observation and observability arises from a suggestion made by several philosophers of science including van Fraassen, Shapere and Achinstein. The suggestion is to do less philosophizing over the nature of observability and instead look to the relevant sciences to see how they classify observability. Van Fraassen puts it this way, "To find the limits of what is observable in the world described by theory T we must inquire into T itself, and the theories used as auxiliaries in the testing and application of T" (1980, 57). My plan is to take this idea seriously and inquire into several theories T, staying primarily within the larger realm of physics, to analyze claims of observability. This will require more than simply collecting instances of the scientists' use of the words "observation" and "observable". It is more than a common-language style analysis which is appropriate here. To get at the informational content of observability claims, that is, to understand what is being claimed about information of the physical world, some general guidelines by which to evaluate observation claims are needed. Science is to be allowed to speak for itself on the issues of observation and observability, but not without some general scheme for interpreting what it says.

The plan of exposition then is the following. The first chapter is devoted to a historical outline of recent philosophical work done toward clarifying the concepts of observation and observability in science. The history begins with Carnap's (1936) statement of logical positivism in "Testability and Meaning", and concludes with Fodor (1984), "Observation Reconsidered". The intent of the quick historical review is two-fold. For one, it is a good way to get a feel for the topic, the nature of the difficulties, and perhaps to spot some unjustified simplifications and conflated distinctions which have led to entanglement and confusion. Second, my plan is to use the exposition of previous work as a forum for pointing out several important distinctions which must be recognized and respected in any productive study of observation and observability. One of those distinctions is between observation and observability themselves. It has not always been kept sufficiently clear that the concept of observability has two parts, one of which is observation. To be clear on observability one must be clear on what it is to observe and, given what is observ*ed*, what can be counted as observ*able*, that is, what is the nature of the modality in observability.

The second chapter presents the general framework through which the examples are to be analyzed. The difficult part in developing such a framework is in knowing how much structure to include such that there is enough to make sense of the examples but not too much as to compromise the empirical nature of the study. My solution to this difficulty has been to compose the framework and study the examples concurrently.

So, though the guidelines in the second chapter appear before the examples in the third, the former were shaped with some influence from the latter. The guidelines amount to little beyond a precise method of realizing the suggestion to allow physical theory to adjudicate matters of observability. The general scheme is one of evaluating observability in terms of physical interaction and exchange of information, a scheme which I come to abbreviate as the interaction-information account of observability. Physical science is expert in both these matters and observabilty can be described in terms of what can interact in an informationally correlated way. Then case by case one wants to analyze the depth and independence of the theorectical description of interaction and the conveyance of information.

The third chapter is the longest. It contains the several examples of observation and observability claims and as such is the presentation of the data for empirical study. The census of examples is divided into four parts. The first is of examples of things unobservable in principle. Something is unobservable in principle if the physical theory which describes the entity disallows its being observed. Two such entites are discussed, objects beyond an event horizon, and single quarks. The second category of examples will include entities which are unperceivable because of the circumstances of the observer. Call these things unperceivable in fact. The fact that human scientists are the size that they are and have eyes sensitive to certain radiation means that such things as atoms and neutrinos cannot be observed in the same way that a chair or a sunrise can be observed. The important difference between these things which are unperceivable in fact and the previous group of things unobservable in principle is that there is no physical law which precludes observation in the case of the former. Nature itself is blind to the color of a single quark, but it is certainly aware (at least in theory) of neutrinos, electric charge, and the like. These latter things interact with their surroundings such that information of their existence and characteristics can be conveyed. In other words, news of these things which are unperceivable in fact can reach the scientific observer through interactions between the entity in question and something else which is observable. The interesting questions to ask in these cases will involve the understanding and influence of the mediating events between the unobservable entity and the observer's knowledge of it. Four examples of things unperceivable in fact are presented, the quantum mechanical state function, an individual photon, something seen with an electron microscope, and a tectonic plate.

The third categoy of examples is labelled as things perceivable. These are entities which interact with the human observer in a way such that information of their existence and characteristics is conveyed. Here there are three examples, thermodynamic heat, acceleration, and a bubble chamber track. This sort of example will be particularly influential as to what it means to observe.

The fourth category of examples is distinguished by different standards than are the first three. It is intended to describe the observability characteristics of entities which were at one time important features in important theories but have since fallen from grace. The handiest example is the ether, or more accurately, the ethers, since many imponderable substrata have come and gone in the history of science. There is something to be learned about observability in general by understanding the observability status of the ethers. The last example then will be of the ethers, with particular attention paid to the elctromagnetic ether and caloric fluid.

Chapter four is a discussion of scientific realism and the implications of the interaction-information account of observability for that issue. The results of the study of observability are summarized and evaluated for epistemic significance. These results are then allowed to speak on the issue of scientific realism.

The work closes with a brief summary in which no new ideas or arguments are presented. The main points of the four chapters are collected and restated; that's all.

CHAPTER 1

HISTORY

1. CARNAP (1936)

Distinguishing the observable from the unobservable is clearly important to an empiricist philosophy of science. The role of the distinction is no more clearly put than by the logical positivists. As presented by Carnap, the positivists' position associates scientific meaningfulness with observability, and the association is carried out in terms of the language of science. Sentences are to be regarded as scientifically meaningful only if the constituent predicates can be linked to observational predicates, or are themselves observational predicates. (Carnap's phrase is "observable predicates" rather than observational predicates as used here. The latter expression is preferable in light of an important distinction between entities which are observable or not, and terms, which are observational or not. The details and impact of this distinction will be presented in their historical place.) Carnap spends much of his time in clarifying the nature of the link between observational predicates and the terms of meaningful sentences, but for appreciating his conception of the observational-nonobservational distinction, it is more appropriate here to expand on his idea of an observational predicate. The domain of such an expansion must be in the language of science rather than the world which science studies. Observability, as Carnap presents it, is a characteristic of the terms scientists use, rather than the entities scientists study.

In the important work (1936) on the link between observability and meaningfulness, the term "observable" is an unanalyzed starting point of the theory. Describing the nature of observability, he feels, is a task for psychology, the product, that is, of the "the behavioristic theory of language" (1936, 63). Without pursuing the details of such a theory, Carnap proposes a rough explanation of what would count as an observable predicate. As a common sensical speculation on the psychological results, he suggests, "A predicate 'P' of a language L is called observable for an organism (e.g. a person) N, if, for suitable arguments e.g. 'b', N is able under suitable circumstances to come to a decision with the help of few observations about a full sentence, say 'P(b)', i.e. to a confirmation of either 'P(b)' or '-

P(b)' of such a high degree that he will either accept or reject 'P(b)'" (1936, 63).

Carnap does not claim to be doing careful psychology and so this characterization of observability is intended to be neither exact nor immutable. He admits that it is a vague definition and that the dichotomy it draws between observational and nonobservational predicates is not sharp. It is, he acknowledges, a somewhat arbitrarily placed line in a continuum of observability. But this inexactness does not indicate that the distinction is unworkable or unimportant. One can imagine a Wittgensteinian discussion to demonstrate that vagueness does not make a concept useless. The famous example of the concept of a game, a concept of inexact definition, but a meaningful, useful concept nonetheless, is intended to represent the general case of nebulous concepts. Though there is no specific feature common to all games, no exact description of the essential feature of games, we do not discard the concept as useless or meaningless. That Carnap's description of the observable is somewhat loose and imprecise does not indicate that it is useless or unimportant.

Most cases of predicates used in science will, Carnap assumes, be clearly on one side or the other of the observability demarcation. The predicate "red", for example, is clearly observational by Carnap's criterion of one's ability to come, with only a few observations, to a decision about the appropriateness of a particular use of the term. On the other hand, a term which describes a reading taken from a mechanical apparatus, an ammeter, say, is not observational. This follows from the description of observability since one needs separate observations to check the properties of the apparatus before one can justifiably decide on the appropriate use of the reading. People can be trained to use an ammeter to draw conclusions about the strength of an electric field, and to come to those conclusions quickly. But one must consider the full story of accepting those conclusions. Consider, for example, the procedures necessary to establish a connection between the ammeter and the electric field, and to calibrate the ammeter for accuracy. One can be trained to quickly notice the strength of the electric field by looking at the ammeter, but the epistemically significant warrant for accepting sentences about the field requires information from a longer list of observations. In this way, most examples of scientific predicates can be classified as observational or not, and Carnap's rough distinction can be said to be useful.

There is another dimension of vagueness in Carnap's characterization of observability, a vagueness which results from the characterization being relative to an organism. Even restricting the organisms to persons, one must further select which human sensual abilities are to be relevant to determining observability. Since one wants a science which is meaningful to a community and not simply to a particular individual, the criterion of observability must be standard to a community. Hence one must specify which abilities such as sight, hearing, etc., are relevant to forming this community. Carnap describes this selection as a matter of pragmatic convention. One must choose the level of visual acuity appropriate to base a standard of observability and choose what observational handicaps to ignore in assembling the linguistic community which decides the observability of predicates.

It is important to notice that the term "observation" appears in Carnap's rough outline of observability. That is, Carnap is describing the observable in terms of the observed, and this points out two aspects of the issue of observability. There is the question of the nature of observation, what it means to observe something, and there is the question of modality, what it means to be observ*able*. Carnap's treatment deals exclusively with the modality. Assuming one knows what it is to make an observation, one can classify the observ*able* in terms of "suitable circumstances" and the relative few observations required to reach a decision.

Carnap builds a great deal on this foundation of classifying observational predicates, yet the classification is never more carefully described than is quoted above. Instead it is assumed that a careful observability distinction can be drawn through a psychological study and that those cases which are now clearly of observational predicates will always be classified as observable. Carnap's final draft of positivism (1956) therefore always uses a distinction between observational language and theoretical language without mention of exactly how the distinction is made or demonstration that it is a legitimate distinction to assume.

2. MAXWELL (1962)

Grover Maxwell's well known response in "The Ontological Status of Theoretical Entities" to the postivists' use of the observability dichotomy serves not only as a presentation of opposing views, but also to sharpen the issue itself and thereby assist future efforts. He is careful, for example, to distinguish analyses of the language of science from analyses of the entitles described by science. Even if the scientific language could, as Carnap suggests, be divided into disjoint sets of observational terms and theoretical terms, this does not imply that the collection of entities that science describes can be similarly divided. Nor does elimination of theoretical terms from the language eliminate reference to theoretical (that is, using Maxwell's implied identity, unobservable) entities. If an electron, for example, could be described strictly in terms of its observable manifestations, using only terms from an observational vocabulary, there is still, according to Maxwell, the notion of the entity, the electron, which functions in explanation and organizing the phenomena. Reference to an electron couched in observational language is nonetheless reference to an unobservable (at least for the present) entity. The point is that analyzing the language is not analyzing what is in the world, and it is to the world of entities and its dichotomy into observable and theoretical entities that one must look for an analysis with ontological significance.

Focusing then on entities rather than terms, Maxwell attempts to eliminate the ontological relevance of the observational-theoretical dichotomy. He describes a continuum in degree of immediacy of observability of entities from things seen with the naked eye, through things seen with eye glasses, windows, and binoculars, to things seen through a microscope. The extent of this list is a function of time and technology. As more and better devices for viewing the world are introduced, more entities appear in the continuum, as, for example, microbes joined with the invention of the microscope. To answer the question "what can count as an observation?" one is prompted to draw a line somewhere in the list, a line of demarcation between what is observable and what is not. But Maxwell describes the shades of immediacy of observation with sufficient detail to indicate that there is no ontologically important criterion which indicates where to draw the line. He fills in the gaps between modes of getting information from the world, gaps which may have seemed like natural divisions indicating distinctions in the warrant for belief about the world. Surely scientists are allowed their eye glasses in making observations, and these are not different in kind from binoculars, or a microscope, or, Maxwell supposes, whatever is the next viewing device to be supplied by the engineers. Any line drawn in this sequence is problematically arbitrary. It is problematic because there is nothing in the world itself which indicates where it ought to be drawn. So insofar as observability depends on immediacy, that is, lack of paraphenalia, demarcating observable from unobservable entities can be only an arbitrary dichotomy.

Maxwell pursues the suggestion, "that it is theory, and thus science itself, which tells us what is or is not, in this sense, observable" (1962, 11). If there are to be genuinely unobservable entities, things whose unobservability is not dependent on technology or time, these must be entities which are "unobservable in principle" in the sense that the theory which describes them, together with theories of human perception, preclude their being observed. In this way, science itself would answer, in a nonarbitrary way, for the modality of observation. But such a classification is, Maxwell argues, both empty and inconsistent with the empiricists' intended use of the observability status. Maxwell supports the former claim by pointing out that if there is an entity x that is unobservable in principle, then the science must include at least one statement about x, namely, "x is unobservable in principle". But if "x" is not an observation term then the statement is, by empiricists' standards meaningless, and the science seems unable to make claims about unobservability. Yet in making this argument, Maxwell ignores his own advice to focus on entities rather than terms. By his own reasoning, the inability of the theory to use nonobservational terms does not preclude it from referring to unobservable entities. There is nothing to prevent the empiricist from employing a notion of unobservability based on entities rather than terms. Van Fraassen's criterion (discussed later in this chapter) is just such a formulation.

There is still the second of Maxwell's accusations against the notion of unobservability in principle, that the classification is, in any ontologically important sense, empty. The criteria of principled unobservability involve both the limits of human perception and the properties of the entity in question as described by current science. Neither of these is logically immune from changing and with it changing the extension of the unobservable-in-principle classification. Regarding human perceptual abilities, it may be the case that humans cannot now perceive properties of individual electrons, using Maxwell's example, but this does not exclude the possibility that someday,, through mutations, humans will have that ability. Since the criterion of principled unobservability is vulnerable to that kind of change, it cannot be a proper criterion for an ontologically relevant dichotomy. It is possible, that is, for human observational acuity to change in time, but is it not possible for an observability criterion, insofar as it is to have ontological implications, to change in time. Hence, the one cannot be used as a basis for the other.

Not only might the biology of human perception change, but we might change our beliefs about the laws which describe the entity in question, the principles of the unobservable-in-principle. The quantum theory now describes the properties of microparticles as being unobservable. One cannot bounce electromagnetic radiation at visible wavelengths off an electron to get information of the electron's position. But this description could change. The quantum theory might not have things quite right, and in an altered version, or in the theory which may replace the quantum theory, microparticles may be described as observable. What Maxwell is doing here is distinguishing the physical possiblity of observability from logical possibility. Scientific laws will adjudicate the former, but our list of accepted scientific laws is not timeless as the grounding of an ontologically forceful distinction must be. It is the logical possibility of observation that Maxwell insists be used for ontological application.

What Maxwell seems to be saying is that the observability claims of science cannot be trusted with ontological significance because the science may change or be discarded. But this is a curious step in an argument for scientific realism, which is Maxwell's desired goal, because it is a blatantly anti-realist interpretation of science with regard to observability claims. He will claim that for other kinds of scientific statements there is a fact of the matter and we can look to those claims for ontological information. But with observability claims there is no fact; what science describes as unobservable one day it may describe as observable the next. There does not seem to be anything in the world which can settle these issues of observability once and for all. The realist Maxwell does not explain why realism fails with this particular kind of statement.

In the spirit of this antirealist interpretation of observability claims, Maxwell suggests that demarcating the observable from the unobservable is an epistemically useful step in science but one with no ontological clout. To point out its usefulness he returns to the analysis of language, but instead of classifying individual terms as observational or not, as Carnap proposed, Maxwell regards whole sentences as the proper units of observability. The measure of observability in this pragmatic sense of the classification is a sentence being "quickly decidable" (1962, 13). The classification of an observation term then follows derivately as, "a descriptive (nonlogical) term which may occur in a quickly decidable sentence" (1962, 13). Since quick-decidability is linked to the state of technological advance and to theoretical understanding, this criterion of observability is more accurately a measure of what is observable-at-the-present-time. Maxwell is clear in his admission that such a concept is useful in science, in particular as a basis of confirmation. And he is equally clear that the classification is no more than useful. It does not refer to some property of the entities themselves.

3. PUTNAM (1962)

Putnam's contribution to the observational-theorectical dichotomy issue (I use Maxwell's title for the debate at this point as it is the issue as Putnam confronted it) is not to refute any particular description of observability or to give any of his own. His efforts are in clarifying the debate by carefully distinguishing issues which had previously been conflated.

The observability question as Putnam found it in 1962 held a confusion of two issues. He points out, in what is a tremendous help to understanding this observability business, that there are really two separate dichotomies at issue. Focussing on the classification of terms, one can distinguish terms as they are theoretical or not, or one can draw a dichotomy of terms insofar as they are observation terms or not. The two classifications are not coextensive, nor should we expect them to be. Being theoretical describes the origin of a term, namely that it is derived from a theory. It does not indicate one way or another whether the term is observational. The term "satellite", to relay Putnam's example, is a theoretical term, since what it is to be a satellite follows from the theory of gravitation and orbits in central force fields. But "satellite" is also an observational term since it describes something which is accessible to perception. The criterion of observational terms does not depend on the term's relation to theory or to the origin of the term. In depends on our perceptual acuity. There is no observational-theoretical distinction; rather there are two separate dichotomies to consider, an observational-nonobservational distinction and a theoretical-nontheoretical distinction.

Furthermore, Putnam points out with examples, there is no fast link between the classification of observation term and observable thing. An observation term can apply to an unobservable thing, as, for example, in the phrase, "red corpuscle of light". "Red" is surely an observation term, yet corpuscles of light are not observable things, the point being that any useful discussion of observability is going to have to be explicit as to whether it is of language or of things. Conclusions in the one domain do not ground conclusions in the other. The message is much like Maxwell's claim that a scientific language purged of nonobservational terms is not necessarily a science without unobservable entities. Putnam's presentation however, is the more blunt and forceful.

While Putnam is clear in distinguishing terms from things, his example to this effect muddles another distinction which is equally important. The example of the red corpuscle of light mixes observability claims of things and properties. That "red" is an observation term is clear enough, but so is it clear that red is an observable property. Likewise one can agree that corpuscles of light are unobservable, and that "corpuscles of light" is a nonobservational term. To thwart the temptation to think that "corpuscle of light" is observational because "corpuscle" and "light" are, the appropriate unobservable entity can be referred to with a single term such as "lightquantum" which is nonobservational. This switch is analogous to replacing "corpuscle of matter" with "atom" to explicate the nonobservational nature of the term. With the distinction between property and thing made clear, there seems to be agreement between observational term ("red") and observable property, and between nonobservational term ("lightquantum") and unobservable entity. The point is that one needs to discuss not only the observability of things but also of properties. This is another distinction which will clarify the discussion.

Putnam's discussion and the examples it relies on assume that the several dichotomies mentioned can be drawn. There is a distinction between observable and unobservable things and though he does not indicate just how the line is to be drawn, Putnam presupposes the distinction to be in place and workable. Corpuscles of light, after all, are clearly unobservable things. There is also the distinction between observational and nonobservational terms. And thirdly, Putnam points out a distinction between theoretical and nontheoretical terms. Putnam's achievement is not to explain how to make these distinctions; he must assume we already know how to do this if his examples are to serve their purpose. His point is rather that the two linguistic dichotomies are not coextensive and that neither can be assumed to be information about the dichotomy of things.

Putnam has exposed a greater complication in the issue than had been appreciated earlier. But a sensitivity to this complication can make the explication of observability the easier. The task is simplified insofar as one can concentrate on one distinction in separation from the others. The goal of the task thus becomes clarified. But for all this, one still finds philosophers pursuing the observation-theoretic distinction.

4. HANSON (1958) and KUHN (1962)

So far the focus of the arguments surveyed has been primarily on the modality of observability. Given what is observed, how does one distinguish what is observable? But it is also important to know the recent history of the treatment of the concept of observation itself. To this end, consider the influence of Kuhn and Hanson on the understanding of observation. Kuhn's and Hanson's ideas can be discussed in tandem since they argue along similar paths to similar conclusions. Their theses can both be traced, I think, to a Wittgensteinian influence. In the final analysis their response to Carnap's initial invocation of an observability dichotomy will be the antithesis of that made by Maxwell. Where Maxwell argued that any term could be regarded as observational, the Hanson and Kuhn claim is that no scientific term is purely observational.

Hanson's argument explicitly, and Kuhn's implicitly, borrow from Wittgenstein's analysis of the concept of seeing-as. As with Wittgenstein, the analysis is in terms of the language, the circumstances and meaning of an observation claim. In explaining phenomena like the duck-rabbit, where one can claim to see the figure as a duck or see it as a rabbit, Wittgenstein indicates that seeing and observing are concept-dependent in the sense that what one sees depends on one's conceptual repertoire. One can see the figure as x only if one has a concept of an x. And it is not that all manner of seeing is seeing-as, but that in the understanding of the nature of seeing-as is the demonstration of the conceptual nature of all seeing. This is because the difference between seeing-as, such as seeing the ambiguous drawing as a duck, and seeing, such as seeing a sunrise, is not a difference in the act of observation but only in the number of applicable concepts. An observation is seeing-as just in case one has more than one concept to associate with the stimulus. A case of seeing involves the same experience of the observer but in the circumstance of a single relevant concept. The point of Wittgenstein's comments on seeing-as, and Hanson's reissue of the same argument, is to point out that all manner of seeing (observing) is sensitive to the observer's conceptual background. And it is not that observers differently equipped interpret the stimuli differently. The differences are there before any process of interpretation. It is that what one sees depends both on what one looks at and on one's conceptual background.

The ramifications of this are most dramatically put in the following way. People with different conceptual backgrounds (different paradigms, for Kuhn, and different descriptive theories, for Hanson) can confront identical stimuli and see (not interpret) different things. If Ptolemy and Kepler could be brought to stand together at dawn and face the east, where one saw the sun rising the other would see the horizon turning away from the sun. If a four year old child claims to have observed a meson shower, we deny it on the ground that the child hasn't the theoretical background for such an observation. He or she could not play the language-game which includes "meson", that is, is not versed in the paradigm of particle physics, that is, is not a member of the tribe.

This analysis of the concept of seeing-as and its relation to the concept of seeing is the motivation for the slogan, "theory-laden observation" as coined by Hanson. And since observations are the basic data for our understanding the world, that is, they are the final authority for what we can say of the world, there is no further criterion by which to check the observation to know what is really there. Wittgenstein brings this up by asking, in seeing x as ____, what is x? He dismisses this as a bad question, bad because it is unanswerable. No mode of observing is free of links to theoretical background. If it is not some scientific theory which is influential in the observation then at least there is some more common, everyday concept like ducks or rabbits to which the seeing is tied.

The most important result of this quick survey of the theory dependence of observation is the realization that the concept of observation cannot be taken for granted. There are two aspects to observability, the nature of observation itself and the modality. This means that there are two issues of theory-ladenness, theory-laden observation, as Hanson and Kuhn describe, and a theory-laden dichotomy. The theories we believe may influence what we see, what we make of the stimulus, but they may also influence how we evaluate observability, how we distinguish the observable from the unobservable. This latter point was one of Maxwell's, that what we classify as unobservable is a function of what theories we believe. The dichotomy is theory-relative. But the Hanson-Kuhn point is that each act of observation is itself theory-relative. One can consistently argue for theoretical relativity in observability distinctions but argue against theory-laden observation. Theory-laden observation, in other words, should not be confused with theory-laden observability.

5. ACHINSTEIN (1968)

Like Putnam (1962), Achinstein in *Concepts of Science* is careful to separate the different dichotomies which are sometimes conflated into a spurious observation-theoretic distinction. One may regard the language of science as being divided into disjoint vocabularies, one theoretical, the other nontheoretical, or alterantively divided into observational and nonobservational vocabularies, but the different criteria classify the terms differently. What is theoretical is not necessarily nonobservational. "Satellite", recall, is an example in support of this claim. Achinstein adds other examples of theoretical terms which are also arguably observational. Such terms as "electron" and "atom" are clearly theoretical. Achinstein claims they are also observational.

He does this by arguing for an extremely liberal interpretation of what it is to observe something. His rather extended use of the verb "to observe" includes perceptual activities in which the object in question is hidden or invisible and only the effects of its presence can be seen. In this sense, an electron is observed by the track it leaves in a bubble chamber, and temperature is observed by its effect on the length of a column of mercury. This is a legitimate extension of the concept of observation, he claims, because it accords with the way scientists speak and practice.

Awareness of the scientist's conception of observation shows that the classification of observability is relative to particular contexts. What it means to claim that x is observable depends on the circumstances under which the claim is made. It depends, for example, on the background theories in place. Given a theory of charged particle ionization of super heated liquids, electrons are observable with a bubble chamber. But without such a theory or some other in its place to account for the tracks, electrons are unobservable in this sense. Achinstein intends a great deal of flexibility to be used in applying the classification of observability. That is, one may require that the object be directly visible or that only signs of the object be visible, the different requirements being applicable in different scientific circumstances. The point is that the proper question to ask of the observability of an object x is not, "Is x observable?" but "In what way is x observable?" The practice of science admits of a variety of methods of observation and it is this latter approach which can reveal the context, assumptions, standards, and methods behind observability claims. The informative exercise is not to simply check off the items of science as observable or not, for this imposes an artificial simplicity on the scientific practice of observation. It is more productive to illustrate the "particular methods of observation" in each case.

I regard this as a very good suggestion. But Achinstein's response to his own suggestion is somewhat disappointing. There is a great deal to be learned by analyzing the details of the scientists' observability claims, but one cannot simply take the scientists at their word. Achinstein's liberal notion of observability comes from an uncritical common-language (common scientifc language) analysis by which what the scientist says is observation is therefore a case of observation. He presents numerous examples of the use by scientists of the concept of observation without carefully explaining what warrants the use and without thoroughly investigating the origin and manipulation of the information. Unlike Maxwell's suggestion to let the theories of science decide observability, Achinstein's approach seems to be one of letting the words and action of the scientists themselves decide the issue.

6. VAN FRAASSEN (1980)

Van Fraassen is clear in *The Scientifc Image* that his interest, and the important aspect in the question of observability, is the dichotomy between observable and unobservable entities. He concentrates on things rather than words. To support his antirealism, van Fraassen must reject the loose, contextual notion of observability as advocated by Achinstein, and return to some form of important, univocal dichotomy between what is observable and what is not. And as he is concerned with entities rather than terms, he can concede the theoretical-nontheoretical dichotomy of language without compromising his position. That is, all of the terms in a scientific language may be theoretically influenced, or, using Hanson's phrase, theory-laden, such that language cannot be divided into theoretical and nontheoretical vocabularies. But this distinction, or lack of it, does not, as has been argued before, extend to the observability distinction between *things*. And the interesting question to ask of science, the question with ontological significance, is of the entities and whether or not they are observable.

Van Fraassen then presents a characterization of observability in terms of what is observed. "X is observable if there are circumstances which are such that, if x is present to us under those circumstances, then we observe it" (1980, 16). This description is intended to solve the modality issue but not to answer for the nature of observation, for surely we must already know what it is to "observe it" for this criterion of observability to be useful. The modality then depends on the possibility of the "right circumstances" and the powers of human perception. There is a difference by this criterion between what appears through a telescope and what appears through a microscope. One could put down the telescope and travel by rocket out to the distant moon or whatever, and see it with the unaided eye. These are the right circumstances. But the analogous trip cannot be made to see the microbe which appeared in the microscope. The world is such that there is no place one can stand, no circumstances such that the microbe is visible to the naked eye. The moon is observable; the microbe is unobservable. The criterion is based on the human body as perceiver, and on the physical possibilities of interaction between observer and object. The modality is one of physical possibility rather than logical possibility, and as such it will be physical laws, scientific laws, which must be consulted to adjudicate observability.

This approach to observability is, not surprisingly, similar to Carnap's in several respects. It is admitted by both, for example, that there is a continuum of what is detectable and that the characterization provides only a vague line of demarcation. But as long as there are clear cases of observables, the moon, say, and clear cases of unobservables, such as a neutrino, the distinction is useful. The Carnap and van Fraassen criteria are alike also in that they are relative to a community. That is, observability is observability for a particular group, and the classification is a function of the group's ability to receive stimulus and to its epistemic (van Fraassen) or linguistic (Carnap) repertoire. The relevant community for scientific observerability is the community of humans, and the relevant classification then is "observable-to-us" (van Fraassen 1980, 19).

As has already been pointed out, both van Fraassen and Carnap invoke an unanalyzed notion of what it is to observe. They start with this primitive observation and get right to the modality. Van Fraassen though, does expand briefly on the concept of observation before using it. The mode of observing that is relevant to the question of observability in science, he says, is in the form of "observing x" rather than "observing *that* x is p". The information "that x is p" is not information which one can get through perception and hence is not relevant to the question of observability. But this dictum is stated by van Fraassen without sufficient argument, and the brevity of his treatment of observation is a weakness in his presentation of the concept of observability.

To see that the concept of observation is more complicated than van Fraassen lets on, consider the example of photons and the problem it raises for van Fraassen's case, as pointed out by Foss (1984). Do we observe photons? Presumably, van Fraassen would want to say that we do not, to avoid the risk that other microparticles might follow. But if we do not observe photons, what is it about the interaction between photons and the human perceptual organs that is missing? There must be something more to observation than being presented with the object and registering an awareness of its presence. The experience must be recognizably *of* photons if it is to count as observation. That seems to imply that it must convey information of the form "that x is a photon" or "that a photon is present".

The point is not that it is seeing-that which must be the basis of scientific observation, although this may turn out to be true. Rather, it is simply that a clear explanation of observing is a necessary part of explaining observability. By leaving the details of the nature of observation uncertain, van Fraassen leaves the details of observability uncertain. And in particular, the observability of photons is left in doubt. It will not do, either, to shrug and say that the case of photons is vague. It will not do because photons are too important. They are representatives of elementary particles and leaving their observability vauge invites an epidemic of unclarity. They are also important as playing a key role in the business of observation itself. If the task is to understand observability then one is obliged to understand the observability of the photon.

Since the modality of observability is based in physical possibilities, that is, whether it is physically possible, given the limiations of the human body, to achieve the right circumstances for observation, discussions of observability must be left to physical theory. Whether x is observable or not is not an issue to be answered a priori by the philosopher. It is an empirical issue, "a fact disclosed by theory" (1980, 57). In other words, to see which objects described by a theory are observable and which are unobservable, one must consult the theory itself, together with the auxiliary theories which are used in doing relevant experiments. The theory itself must distinguish some of the objects in its domain as observable and others as unobservable. And it is then belief in the claims about the observables, with enforced agnosticism regarding claims about the unobservables, which is the battle cry of van Fraassen's constructive empiricism.

For constructive empiricism to get off the ground, the theoretical statements of the form "x is observable" or "x is unobservable" must be interpreted realistically. These must be statements of facts. It is either this or a priori assumption that the property of observability is itself an observable property. But this is not letting the science speak for itself. This is the philosopher telling it what to say.

Allowing the sciences to comment on the nature of observability is clearly a promising move, but it seems prone to induce a curious reaction. In the hands of a realist, Grover Maxwell, the observability claims of science are given an antirealist reading. In the hands of an antirealist, Bas van Fraassen, observability statements are interpreted realistically.

7. SHAPERE (1982)

Shapere's "The Concept of Observation in Science and Philosophy" is a careful realization of the plan to let science itself report on the classification of observability. He uses the relevant physical theories to describe not only the modality, what is physically possible, but also to clarify the concept of observation, that is, to make precise what it is to observe.

Emphasizing that it is observation as it is understood and practised by scientists that is at issue, Shapere generalizes the notion beyond ordinary, philosophical bounds. Observation is a subspecies of interaction, where by interaction he means exactly what the physicist would mean by the same term. Observation of x is any interaction between x and a receptor such that some information about x is transferred to the receptor. Human beings are not the only kinds of receptors in Shapere's description of observation. Any device which can interact with the object in question, and interact in an informative way can function as a receptor. In this scientific notion of observation, no human observer need be involved in an act of observation. Geiger counters observe charged particles even when left alone. It is this separation of the human scientist from the act of observing which is the basis of Shapere's argument for the objectivity of science. The thrust is to associate observation with interaction in general, and "it is that process of integration that frees observation from the subjectivity of its philosophical associations and the unreliability of its ordinary ones" (1982, 510).

With observation as interaction, the observation process is amendable to description by physical theory, general laws of physical interaction, together with more specific theory of the source, the object to be observed, and theory of the receptor. The physical laws will describe not only the circumstances under which interaction can take place, the modality of observability, but also will correlate the information of the object with the resulting information in the receptor after the interaction has occurred. By generalizing the notion of the receptor to include inanimate tools as well as human sense organs, the question of observation becomes one of epistemological import rather than perceptual. Observability in this sense no longer hinges on the sensual acuity of the human body, nor on an unambiguous understanding of perception. It hinges instead on an account of the transfer of information from object to receptor, whatever the receptor might be. Shapere explicitly rejects (1982, 509) the suggestion that the information must eventually be perceived by a human scientist. Observation, in his sense, is complete when the information resides in the receptor, whether it is human or machine.

In Shapere's account, both observation and observability are theory dependent. Since one relies on some physical theory to translate the information of the object of observation, the information one gets by observation depends on the theories used. In this sense, observation is theory dependent. Similarly, relying on theories to decide what can interact with a receptor and what information is possible through interaction, makes observability theory dependent as well. But these dependencies are not problematic, that is, they do not compromise the objectivity or rationality of science. This is true because the observation of some object x can be supported by theories which are independent of the theory of x and wholly indifferenet to x's properties and x's existence. In this way, one theory is used to bolster the observation to test another, and the process is a rational one so long as there is no reason to doubt the auxiliary theory which supports the observation. Observation is incorrigibly theory dependent but care can be taken that it depends only on the best.

Shapere distinguishes between direct and indirect observation. An observation of x is direct if the physical medium which carries the information of x actually comes from x or interacts with x, and itself reaches the receptor. Contrast this with indirect observation of x in which the medium which intereacts with x then interacts with something else and passes the information on to what it interacts with and this second medium then either reaches the receptor or interacts with yet another medium, and so on. Shapere's example of direct observation is observation of the interior of the sun using solar generated neutrinos as the medium of information. The same neutrinos which leave the solar interior interact with the receptor, or so the theory goes. Another example of direct observation in Shapere's sense would be observation through a miscroscope where the photons which interact with the specimen also interact with the receptor, the eye. Shapere's example of indirect observation is again observation of the interior of the sun, but this time using photons as the medium of information. Virtually none of the photons generated in the solar interior escape to reach a receptor on the earth. Instead, they die in collision, pasing some information on to other photons who travel similarly short distances. Any photons which emerge may carry some information on the original, interior-produced photons, but they do not get the information directly from the solar interior.

The distinction between direct and indirect observation is sensitive to where one draws the boundary of the receptor. In the neutrino example, Shapere describes the receptor as everything from the tank of cleaning fluid in which the neutrinos are supposed to interact, through the chemical apparatus and the particle counters. The neutrino reaches the tank, so it reaches the receptor. But the neutrino does not reach the particle counter, it sends positrons from decaying argon in its stead. And it is not clear why one cannot extend the receptor in the indirect cases to include not only the telescope and photon analyzers on the earth, but also the outer material of the sun, using it in a way analogous to the tank of cleaning fluid. Now the interior photons do reach the receptor, suitably enlarged. The moral that I suggest follows from this ambiguity is that the dichotomy of direct-indirect observation is artificial. This does not mean that the experiences of the information media, and the number of media changes are unimportant. It is still an indicative measure of directness of observation, and it is valuable as just that, a measure. Call one observation more direct than another when the first involves fewer changes in the medium than the second. But without a clear and nonarbitrary description of what counts as a proper receptor, the dichotomy of direct-indirect observation can only be arbitrary.

8. HACKING (1983)

In *Representing and Intervening*, Ian Hacking distinguishes between two aspects of the scientist's contact with the world, observation and experiment. Experiments are long term processes in which the person exerts some influence on some part of the world, intervenes, for examples, by building machines to move electrons and steer them into collisions. The world in turn responds to influence the person by display of tracks in the bubble chamber. The observation is only a brief part of the experiment and it is not necessarily the most indicative of what there is in the world.

With regard to observation, Hacking suggests that the claim of theory-ladenness is over done. If it is a claim that any belief, even basic, common sense beliefs such as that the intermediate air does not distort a view, is a theory and can bring theoretical influence into the process of observation, then it is simply "trifling". But he points out that the influence of a particular theory can be limited in a particular way such that one can observe at least in a way that is independent of that particular theory. (This ability to block particular theoretical influences anticipates what we will see in Fodor's reconsideration of observation.)

Hacking supports and illuminates his claims about observation with a detailed example of observing through a microscope. The result of the example is that, van Fraassen to the contrary, one does see (observe) through a microscope, and furthermore, the observation is independent of the optical theory of the microscope. Technicians can be trained to notice particular features in a microscope even though they have no appreciation of the theory which describes how a microscope works. In fact, considering how sophisticated and diverse the theory of microscope optics is, almost no one who uses a microscope really has the theory in mind sufficiently clearly to have it influence their observation. This theory at least must be isolable from that observation. One can believe that what one is seeing in the microscope is an accurate image of the specimen and is not an artifact of the machine even without having a clear understanding of how the machine works. Proof of that is in the history of microscopy in which the theory of how the microscope works changed drastically, from absorption of light to diffraction of light as the relevant phenomenon, with no change in the belief in the veridicality of the image (Hacking 1983, 99).

It is important to point out that Hacking's use of "observation" is idiosyncratic. In places he equates observing with noticing (1983, 180), and one can be said to observe through a microscope in a theory independent way if one can notice features in the image. The beliefs which are then supported by this noticing are beliefs about the accuracy of the image. One "believes the picture" when one trusts the machine to reproduce the image without distortion or artifacts. But this kind of theory independent noticing does not lead to beliefs about *what* is being seen. Knowing what you are seeing does require influence by theory.

Hacking does more than simply say that one can see through a microscope; he presents an argument in the claim's behalf, the "argument of the grid" (1983, 202). It is possible, and in fact routinely done, to draw macroscopically visible grids and reduce them photographically such that the grid is invisible to the naked eye but clearly visible through a microscope. Through the microscope the shrunken grid appears identical to its macroscopic, preshrunken state. It is the ability to manipulate the grid, to blacken one of the squares of a large grid, say, shrink it to microscopic size and see the blackened square in the microscope that convinces Hacking that one must be actually seeing the specimen through the miscroscope. In this sense one has causally affected the microscopic grid and for Hacking, what is manipulable is believable. Where van Fraassen's modality is based on being able to go and see with your own eyes, Hacking's modaility is based on intervening. What you can push, reshape, or maniuplate in any way must be there. Furthermore, if it is not that the microscope is veridcal in its imaging of the grid, it would require a huge coincidence to explain why the reduced grid appears through the microscope exactly as it had to the unaided eye. Such is the abductive argument of the grid.

The example of seeing through a microscope is rounded out by a general description of what it is to see. "When an image is a map of interactions between the specimen and the image of radiation, and the map is a good one, then we are seeing" (1983, 208). A map is a good one if it preserves the structural spatial relationships in the specimen. But observability, Hacking continues, is more or less incidental to the issue of what there really is in the world. It is not the impression that can be made on the scientist which is an expression of existence; rather it is the impression the scientist can make on the world which holds the ontological information. Believe, he counsels, not necessarily in what can be observed, but in what can be interfered with. And here the microscope examples serves his cause well. We can believe in what we see in the microscope because we can poke at it and move it around, shrink it, and generally interfere with its appearance. It is the causal influence we have on the world which is the more important basis for belief than is the causal influence the world has on us.

9. FODOR (1984)

Like Shapere, Fodor has an obvious axe to grind in his examination of the concept of observation, that is, to ground the objectivity and rationality of science. Fodor's reconsideration of observation is interesting because it explicitly points out an important distinction, though at the same time it conflates another.

Fodor points out that one can distinguish the claim that observation is inferential from the claim that observation is theory-laden. And in fact, the first can be granted while the latter is denied. (This is exactly opposit of the Wittgenstein-inspired account in which observation is theory dependent but not inferential.) The case that observation is not theory dependent is based on empirical evidence that inference involved in observation does not have access to the whole of the mind's theoretical library. Fodor's evidence in "Observation Reconsidered" is of observation of the Muller-Lyre illusion where the lines look to be of unequal length even to someone who knows that they are not, and who furthermore knows the theory which explains the illusion. Clearly, any inference in support of this observation cannot be using the theory of the illusion, and try as one might, the inferential process cannot be made to use the theory. This theory at least is inaccessible to this observation. One's theoretical commitments about perception and the Muller-Lyre illusion cannot influence the observation, and in this sense the observation is theory neutral. Case by case, it is "entirely empirical" as to what theories are part of the accessible background of an observation. The crucial question regarding observation in physics is whether physics belongs to the accessible background. Fodor is willing to bet that it does not.

This concept of theory neutral observation is then used as the basis of a theory neutral observability dichotomy. Things or properties are observable only if they are referred to by the terms in accessible background theory (1984, 38). So if particle physics is part of the inaccessible background, and that is an empirical question, then protons are not observable. This classification is not theory dependent in the sense of being changeable with a change in what theories one holds. It is a distinction "in the nature of things".

With this criterion of observability Fodor seems to be running together an observation-nonobservational dichotomy of language, and an observable-unobservable dichotomy of things. The need to separate these issues was demonstrated by Putnam and van Fraassen. Fodor's analysis faces difficulty, for example, with an entity referred to as, "a particle too small to see". The composite term must be observational because none of the constituent terms is in anything but accessible background, or so I assume. But the entity referred to is not observable, at least it should not be classified as observable on this basis.

Furthermore, it is important to point out that Fodor's analysis is of perception but not of beliefs formed from perception. The epistemic functions of deriving knowledge from the senses are not isolated from inaccessible background theory. "Belief, fixation, unlike the fixation of appearances -- what I'm calling observation -- is a *conservative* process; to a first approximation, it uses everything you know" (Fodor 1982, 40). The distinction is important, and it is the epistemologically important process of belief fixation which will be relevant in understanding the nature of scientific knowledge.

10. IN GENERAL

This sketch of the recent history of work done on the concepts of observation and observability is presented not simply to get it straight who said what, but also as a vehicle to show certain important distinctions in and around the issue. If we are to benefit from the history, these distinctions which serve to clarify the question should be respected. I summarize them now.

From Putnam, Achinstein and van Fraassen it is clear that one must be explicit whether it is an observability dichotomy or a theoreticity dichotomy which is at issue. The classifications made on the basis of theoreticity are not necessarily coextensive with those made on the basis of observability. Nor should one expect that they would be, since being theoretical describes a term's semantic relationship to other terms and beliefs, whereas observability is a measure of the manner in which something comes to be known. I should make it clear then that the focus of what follows here will be on observability.

One must also be clear whether it is the language of science or the entities to which the language refers which is to be analyzed. Perhaps the scientific vocabulary can be divided into two separate vocabularies, observational and nonobservational terms. And perhaps the world can be divided into two groups, observable and unobservable entities. But these are independent questions. Success with the first dichotomy does not, as Maxwell pointed out, imply success with the second. The scientist, or anybody else for that matter, can describe unobservable things using observational language. Following van Fraassen's insistence that what count are things and not words, I will be analyzing entities and their observability.

There is also the distinction pointed out by Shapere between a concept of observation which is sensitive to the details of perception and one which focuses on the epistemological import of observation. In studying observation as an epistemic act as I plan to do here, the human sense organs as receptors are generalized to all types of detectors used in science. Observation then becomes, at least in part, a kind of interaction between object and receptor, an interaction in which information of the object is passed to the receptor and eventually, as I want to look at it, to the scientist. The task of an analysis is then to locate and follow the information from specimen to observer and to discover what background knowledge and auxiliary theories are required or influential to translate the information and make sense of the observation. If it is, as Fodor indicates, entirely empirical, then this is what must be done to find out what functions as accessible background. It is time to call in the bets.

And finally, the work will be clarified by being explicit whether it is observation or observability which is on the line. The claim of theory dependence can be discussed for either concept, whether all observations are influenced by theory as opposed to whether a line between what is observable and what is not can be drawn in a way independent of theory. Theory dependent observability for example, does not imply theory dependent observation, and in advocating or refuting the charge of theory dependence it is best to keep the two issues explicitly separate.

There are two steps to describing observability. There is first the question of what it is to observe, a question left unanswered by Carnap and van Fraassen. And there is the question of modality. That is, given what is observed, what then is observable? There seems to be a consensus in the above survey to allow science itself to respond to this question. Letting science speak for itself on the observability issue requires hearing it on many cases since science speaks in many tongues and confronts a diversity of situations. It requires also that one avoid imposing any rigid observability standards with which to evalute the examples. But neither can one simply, uncritically list and listen to all of the uses of the verb "to observe" or all the uses of "observable". Not only would one learn little or nothing from this, but, frankly, the scientist cannot be trusted to always use the terms in an epistemologically careful way. To learn from the examples, to see where the information comes from, where it goes, and what it means when it gets there, the study wants some general guidelines on what to look for and how to trace the flow of information. It wants a general idea of how observability reports can relevantly differ. It is such a general scheme, a set of guidelines for studying observability, that we turn to next.

CHAPTER 2

DIMENSIONS OF OBSERVABILITY

There is an aspect of the practice of philosophy which involves describing some important and frequently used term by specification of necessary and sufficient conditions for the proper use of the term. Much of epistemology, for example, is engaged in articulating an accurate description of "knowledge". And with an accurate description is the implication of accurate and inaccurate, or right and wrong uses of the word. This approach to understanding the concept takes on a normative tint by which some immutable, prescribed meaning of a term must be respected.

I want to avoid this in my treatment of the concepts of observation and observability. That is, I want to avoid conclusions of the form that specifies what ought to be classified as observation, or what ought to be labeled observable. Whatever it is called, I simply want to see what is going on. In a case, for example, of learning about the structure of a cell by placing a specimen on the stage of a microscope and looking through the microscope, it is not so important whether or not we call this an observation, or whether we label the cell as an observable or unobservable, so much as it is important to understand just what is going on with respect to observation in this case. What makes it different from looking at a chair or learning about cell structure by biochemical analysis? Which, if any, account of information about the object is the most reliable? You can call the process anything you want as long as you know and explain what you are talking about.

All this is intended to head off a particular kind of objection to my characterization of observability and my examples of observation reports, objections of the form, "but that's not really *observation*." That is a quibble over terms, and so long as it is clear from the beginning how the terms are to be understood, it is a quibble without substance. The purpose here is to describe the process of observation and the classification of observability where those concepts are given broad interpretation, rather than to prescribe proper interpretations. This way one comes to understand the problems of observation and observability in science, though not to impose solutions. For this kind of study, trust Humpty Dumpty, " ' When I use a word,

'Humpty Dumpty said, in rather a scornful tone, 'it means just what I choose it to mean -- neither more nor less.' " The first step then is to be clear about the use of "observation" and "observability" in this analysis.

1. SCIENCE, KNOWLEDGE, INFORMATION AND OBSERVATION

The focus is on observation in physical science and both "observation" and "observability" should be read with an implied "physical scientific(aly)" in front of them. Science is in the business of gaining knowledge about the physical world, and observation, as it is part of science, must contribute to the knowledge. Observation must be not only a physical event but an epistemic event as well. One expects there to be some aspect of knowledge which is supported by the observation in the sense that the knowlege is possible with the observation but not possible without it, and it must be knowledge of the physical world. Furthermore, knowledge is of facts, not particulars. One knows things of the form "x is P" but not of the form "x" or "P". That is, knowledge is built of information, where the elements of information are facts, "that x is P".

As an initial generalization then, one can say that observation is a manner of getting information of the physical world, from the physical world. It is accomplished with the conveyance of information from the world to the scientist, information like "x is P", where x and P are features of the physical world. Conveyance of information is crucial and a simple event of interaction without information exchange is not interesting insofar as epistemic processes are the topic. Events which might be described as observing x, for example, observing the table, are not epistemically valuable until we know what it is about x that is observed. We might grant as unproblematic that one observes, receives information, that the table is present or that the table is round. But our confidence decreases in the case of observing that the table is composed of molecules. The point is that the important epistemic value of an observation report is in its informational content, and informational content must include not only a report of the object x, but also a report of some property P, the what-it-is-about-x that is observed. Sensory events without information exchange are not epistemically useful and so are not scientifically useful.

This is intended as a part of the explanation of the use I plan for the term "observation", and it is not supposed to claim that this is the proper use or only use. It is just a clarification that this is what we will be talking about. "Observation" refers to an event, or a series of events, which are epistemically efficacious in the sense that information is transferred in the event. "Observability" describes features of the world and applies insofar as they participate in these sorts of events.

2. THE INTERACTION-INFORMATION ACCOUNT

I now describe the basics of the mechanics of observation and the criteria of observability. Care must be taken at this stage not to compromise the empirical nature of this study of observation and observability. This is not the time to draw conclusions hard and fast about the nature of observability because the examples have not yet been presented. Science has not yet spoken, so the evidence is not yet in. At this point only very general guidelines with which to understand the examples of observation reports are called for. And they certainly are called for to avoid an uncritical reading of the scientists' use of terms like "observation" and "observable". What is wanted as a prelude to the evidence is a sketch of observability (and observation) which is appropriate but neutral to the outcome of the two guidelines set so far, that the accounting of observability be information theoretic and that observability be a question whose answer is provided by science itself.

The first constraint motivates an application of observability as a two place predicate. This is because the nature of information specifies the two place form, "that _____ is ___". To respect the informational aspects of observation and observability it will be necessary to assess the observability of an object and a property together, rather than an object or property alone.

The second constraint indicates that the description of the conveyance of information ought to be in terms of physical interaction. Interactions, as described by physics, psychology and other sciences in between, are the ways things get done in the world.

In accordance with these suggestions, a general characterization of observability can be presented and used as a guide for organizing an empirical investigation. It should leave open the possibility that observability may be a question of degree rather than simply a dichotomy:

> The ordered pair < object x, property P > is observable to the extent that there *can* be an *interaction* (or a chain of interactions) between x and an *observing apparatus* such that the *information* "that x is P" is transmitted to the apparatus and eventually conveyed to a human scientist.

I have underlined the troublesome aspects of this characterization and insofar as ambiguous words like "can" and "observing apparatus" are left unanalyzed, this description means very little. The crucial points in observability, after all, depend largely on the nature of the modality and the machinery, or lack of, used in the event. I will remedy these ambiguities by explaining in detail what is meant by the underlined terms. But first I should present the analogous characterization of the concept of observation. This amounts to specifying what it is for an object-property pair to be observed:

> The ordered pair < object x, property P > is observed to the extent that there is an interaction (or a chain of interactions) between x and an observing apparatus such that the information "that x is P" is conveyed to a human scientist.

Derivatively, one can speak of object-observability and property-observability:

> The object x is observable to the extent that there is a property P such that < x, P > is observable. And, the property P is observable to the extent that there is an object x such that < x, P > is observable.

These concepts potentially admit of degrees in the sense that x is more observable than y if more of x's important properties P are such that < x, P > is observable than is the case for y and its important properties. A similar scaling is possible for properties. Since these classifications depend in a straight forward way on the more basic object-property pair observability, it is this latter conept which will concern us here.

Turn now to clarifying the underlined terms in the description of observability. The "can", the modality of observability, and the "interaction" are both concepts which can be handled entirely by the physical theories of the situation, the physics, biology and other sciences of the source or media. This is because physics and the specific physical laws which describe the object and the observer are exactly descriptions of the possibilities and nature of physical interactions between these things. The fundamental laws of physics, for examples, include specification of the circumstances under which elementary interactions occur. For example, electromagnetic interaction can happen when two particles are electrically charged, but not otherwise. The fundamentals of physics also include description of what happens in an interaction, or, put another way, what the interaction is. It involves an exchange of some conserved quantity such as energy or momentum. And it is this account of interaction which must be used to describe observability if that description is in the language of science. Physics describes four basic kinds of interaction (nuclear strong and weak, electromagnetic and gravitation) and in the final analysis these are the only ways that influence can be transmitted in the world. It is only by one of these four interactions that information can be exchanged, whether it is from one object x to another object y, or from x to some machine for detecting x's, or from x to a person.

Whether all information exchange can be accounted for by the physics of elementary forces is a question of the reduction of other sciences, biology, geology and even the rest of physics, to the physics of elementary forces. It is wise to avoid this issue and allow as interactions not only the four elementary interactions but also groups of interactions which are manifested through other domains of science, as well as chains of interactions as mentioned in the above characterizations. For the purpose of studying observability, the classification of groups and chains of interactions should be made on the basis of informational exchange. A concert of interactions may convey some information together which no individual or subgroup contains. This is then an (information-theoretic) group of interactions. Groups of interactions occur when macroscopic objects interact with each other as for example when an opaque object interacts with a beam of light in a way that the information of the object's shape, that the ball is round is conveyed. This information resides not in any one of the object's microparticles but in all of them as a group, and it is carried off, not by a single photon, but by a group of photons. At the observing end, the relevant biological description is of interaction between the group of photons and a part of the eye, itself a group of atoms. The information in this case is communicated not by a single interaction between photon and atom but by a group of interactions. The criteria of what counts as a group of interactions are information theoretic. If several interactions, when regarded together convey information that no interaction does by itself, they are regarded as a group of interactions. It is, for example, a group interaction which informs me of the momentum of a ball that hits me in the head.

The idea of a chain of interaction is also information theoretic, and is no less intuitive. There is a chain of interactions from x to b just in case x interacts with y, and y with z, and z with . . . ., . . . with b, there being as many intermediate interactions as you like. So a chain of interactions (or a chain of groups of interaction) is any temporal sequence of intereactions (or group of interactions) in which each successor event shares a participant with its predecessor. Information can be carried along a chain of interaction such that information of the original event and intermediate events, can arrive at the final event. It is not the case that all the information of the first stage, or even any information of the intermediate stages must always be carried to the end. But the thread which runs the length of the chain and identifies it as a chain of interactions in a sense that is relevant to a study of observables is that some information of the initial object x is contained in each step. Accounting for the flow of information, that is, unpacking the information about x and describing the reliability of its transfer along the chain, is an emperical task to be carried out on a case by case basis.

The picture of observation and observability may begin to look messy and complicated with these concepts of chains and groups of interactions, but they are unavoidable since even the most unassailable cases of observables, the clear cases, must, when evaluated by the relevant physical laws, be described in this or similar language. The physical science account of observing the back of your hand is in terms of interactions first between your hand a a group of photons and then between the photons and pigments in your eye. There must be not only groups for macroscopic observation, there must be chains of groups of interactions as well. Wherever there is a medium of information, that is whenever one uses light to see or sound wave to hear, there are chains of interactions.

The point is that if one is to let science report on the nature of observability and observation, then observation must involve interaction. Interaction between x and an observing apparatus is a necessary condition for observation, but alone it is not sufficient. More conditions will have to be added, conditions relating to the nature of information.

This talk of interaction is not intended as a new proposal of what observation is. It is simply the natural first step to describing the nature of observation and observability in the language of science. It is the beginning for consulting physical theories on the nature of observation and for allowing science to adjudicate observability. Consulting the science of the situation is the way to handle both the description of the interaction and the criteria of possibility for interaction.

Another problematic concept in the characterization of observability (and observation) is that of the observing apparatus. It is in the interest of openmindedness that I consider the use of machines in observation. This is, after all, to be a manner of studying the informational aspects of observation and observability reports (in science), and it would be rash to ignore cases of information routed through machines. It is not the proper use of the term "observable" which is the goal here. It is the more humble hope for an understanding of how the features of the world inform the scientist. The directive for this chapter is to develop a method for describing the conveyance of information from the object to the observer, and for evaluating claims of having received information from the object and about the object. This method is then to be implemented in the next chapter. In this characterization then I should allow the widest possible interpretation of the observing apparatus (or informing apparatus). It can be a human being, or it can be any apparatus which is capable of processing information from its own useful form, the form to which it reacts, into a form which is epistemically useful to humans. A simple thermometer is a good example. It processes information in the form of average kinetic energy of molecules into information in the form of the length of a column of mercury. The key is that the information which goes in must come out in a form which is accessible to the human scientist. The criterion of observability of $< x, P >$ depends on whether such informational form is physically possible. An observation depends on the act actually being consummated in the sense that the scientist has helped herself to the information and it is staged to affect beliefs and knowledge. This is part of what is meant by observation being an epistemic event.

Clearly, a lot hangs on the notion of information, and it is only by clarifying this that a full understanding of the observing apparatus is made clear and that the necessary condition of interaction is augmented with sufficient, epistemic conditions. What is required is an explanation of what it is to have the information that x is P and how that information is conveyed in interaction. To this end I borrow heavily from Dretske's *Knowledge and the Flow of Information,* and adapt his semantic theory of information to the purposes of observation and observability in physical science. The movitvation for a precise description of information is to preclude disagreement over what it is that one observes. If we disagree over whether one can observe that the table is composed of molecules, then we will approach a resolution to this disagreement by invoking a shared notion of information. What information *do* we get by looking at this table?

Information can be roughly identified with a reduction of possibilities. For example, hearing the rain on the windows carries information because it distinguishes among the possible situations that it is raining or that it is not raining. If in fact I live where it is always raining and I know that it always rains, my hearing the rain on the windows carries no *new* information about weather conditions. It does, however, carry what one can call, adapting a Dretske term, redundant information.

Information content of a state of affairs can be specified quite precisely using a modified version of Dretske's definition of "informational content" (1981, 65).

> A state S has the *new information* that x is P if and only if S, k $\vdash$ (x is P), but it is not the case that k $\vdash$ (x is P).

k represents what is already known about the source, that is about x. For example, if in playing the game of peanut-under-the-shell we are initially told that the nut is under one of three shells, A, B, or C, then the situation in which shells A and B are down and shell C is lifted with no peanut revealed contains the information the the peanut is under either A or B. This follows simply from the facts that (x is not under D) plus (x is under A or B or C) imply that (x is under either A or B), but (x is under A or B or C) does not.

If one already knows that x is P, that is if (x is P) $\in$ k, then S cannot convey *new* information that x is P. But for purposes of observation one wants a concept of information that allows observation of what is already known. For this reason I augment the description of new information with one of redundant information:

> If (x is P) $\in$ k, then a state S has the *redundant information* that x is P if and only if S, k ~ (x is P) $\vdash$ (x is P), but it is not the case that k~ (x is P) $\vdash$ (x is P).

Here the expression k ~ (x is P) means the body of knowledge k with the fact "that x is P" removed. I do not claim to be able to specify how this is done. But it is not essential in understanding what redundant information is to specify precisely how to recognize it. Redundant information is nothing more nor less than what would be new information if you did not know it already.

The above description of information is exactly Dretske's except for the following modification. What he calls "information" I call "new information" and then I add the description of redundant information. In place of his "signal r" I use "state S", a rewording he sanctions (1981, 65). Where I describe information in terms of implication $\vdash$, he uses probabilities so that S, k $\vdash$ (x is P) he writes as Prob(x is P | S, k) = 1. My translation, I maintain, captures Dretske's meaning and presents information content in a way which is more amendable to application to physical laws and less prone to misunderstanding. Dealing with logical implication rather than probablities delivers us from difficulties in interpreting the probability theory.

In the characterizations of observability and observation, the relevant concept of information is either new or redundant information.

Information is transferred between states through interaction. The object in state S which has informational content (x is P) interacts with something else, the observing apparatus or some intermediate informational medium, with the result that this latter object is left in a state A which has the information (x is P) whereas it did not have that information before the interaction. The information (x is P) may be nested (this Dretske's term) in other information, such as (y is Q), of the A-state. Information (x is P) is *nomically nested* in the information (y is Q) just in case there is some law of nature which says that if y is Q then x is P. The state defined by the fact that y is Q has informational content that x is P, and any other state which has information that y is Q also has the nested information about x. This is important in following the conveyance of information along a chain of interactions since the form of the information usually changes in interaction and it is only by unpacking what is nested that information of the object of observation is revealed. In a photelectric interaction, for example, the initial photon state may contain the information that the photon has energy = 1 keV (x is P) while the subsequent electron state has the information that the electron has energy = .997 keV (y is Q). Licensed by the physical law describing photo emission of electrons, and a knowledge of the binding energy of the material, the photon state information is nomically nested in the electron state information. The electron carries information about the photon. The path of information proceeds from photon to electron, pictured as (x is P) ⟶ (y is Q), indicating the temporal sequence of interaction. The recovery of information, the reconstruction of S-state information goes the other way, (y is Q)------→ (x is P), where now the dashed arrow indicates a nomic implication. And here is a crucial point where relevant physical theory is invoked to adjudicate matters of observability, since it is the physical theory which must describe and sanction the nomic nesting.

Typically, actual cases of observing involve longer chains of interaction than the one-interaction example above. In these cases the information (that x is P) of the object state may get passed along the chain becoming nested more and more deeply with each interaction. From the source S-state with its information that x is P, through states of the media (y is Q) ⟶ (z is R) ⟶ (v is S) ⟶ . . . to the final observing apparatus A-state (u is T), the information is conveyed. The chain of interaction delivers to the observer the information that x is P if (x is P) is nested in (u is T). The trick to observing < x, P > in this sense, and the focus of evaluating examples of observability claims, is to correlate the A-state information to S-state information.

It would be mistaken to simply identify the flow of information with a propagation of causal influence, because it is possible for one state to be the cause of another without the latter state getting information of the former. Suppose, for example, a state S' with informational content (y is Q) interacts and causes a subsequent state A with information content (z is R). If S' is not the only possible cause of A, that is, if the A-state information (z is R) could also result from interaction of another state S" with informational content (v is S), then A does not get the S'-state (or the S"-state) information since the information (z is R) does not distinguish between possibilities (y is Q) or (v is S). The point is that causal interaction is not sufficient for the conveyance of information.

The transfer of information cannot be identified with deterministic interactions either. Consider a chain of interaction which begins with a state S with information (x is P) which can interact in one of two ways, resulting either in state S' with (y is Q) or S" with (v is S), where there are no laws determining which of the two states will result in any particular case. Both S' and S" carry S-state information (x is P). But suppose further that both S' and S" will interact to produce a new state A with information (z is R). Now the A-state contains S-state information (x is P), but has no information of either intermediate state S' or S". Schematically, the interaction process looks like this:

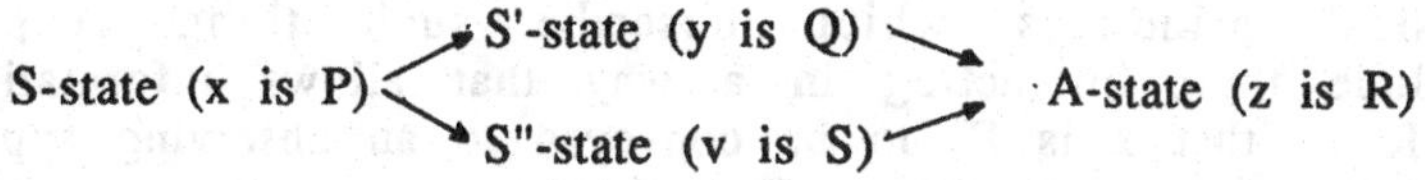

S'-state (y is Q)
S-state (x is P) → → A-state (z is R)
S"-state (v is S)

The information (x is P) is nomically nested in the information (z is R) but no S'- or S"-state information (other than the information that either y is Q or v is S) is nested in (z is R). With an interaction chain such as this the observing apparatus can get information of the object, the S-state, while receving no information about the medium which carries the information.

The basics of this interaction-information account of observation and observability are this: Observability (observation) requires that there can be (is) an interaction chain from the object, the S-state, to the A-state, a human observer or the final state of some appropriate observing apparatus. The A-state can be correlated to the S-state according to nomic relationships between the two, and in this way the observing apparatus has information about the S-state, information which comes from the S-state.

3. DIMENSIONS

This characterization facilitates an evaluation of the observability of < x, P > in terms of several degrees of freedom. The concept of observability is not one dimensional, and any careful account of observability must operate on all the dimensions.

One of those degrees of freedom, the one most likely to be thought of as *the* measure of observability, is the *immediacy* of the S-state to a human observer. Roughly, by this I mean an answer to the question, "does it take some fancy machine to see this thing, or can I see it with my own (unaided) eyes?" Less roughly, by immediacy I mean a description of an object's potential to interact in an informationlly correlated way. The more precise question is, "with what can x interact in an informative way?" There are three kinds of answers to this questions, and I use them as indicators of three possible values of immediacy.

One possible answer is that the object x simply cannot interact with any observing apparatus, whether it is human or otherwise, in an informationally correlated way. I will call such things (with apologies to Maxwell) *unobservable in principle.* It is principled unobservability in the sense that the theories and physical principles which describe such things explicitly preclude their interacting in a way that allows information of the form "that x is P" to be conveyed to an observing apparatus, any observing apparatus. That there are such things, and the details of their classification as being unobservable, will be the topic of the beginning of the following chapter.

A second kind of response to the immediacy question is that the object in question can interact in a correlated way with some nonhuman observing apparatus but cannot so interact with a human sense organ. Call such an object *unperceivable in fact.* This terminology is potentially misleading and I should explain the motivation behind its use. An electron is a good example of something that is unperceivable in fact because it can interact with an observing apparatus, namely a bubble chamber, in an informative way, but cannot so interact with the human observer. But it is a contingent property of the world that the human body does not have bubble chambers as eyes, or as parts of eyes, and that it is not sensitive to electron information in some other way. It is a fact of the situation that the electron can be observed with the aid of machines, and a fact of the perceptual abilities of the body which make such things unobservable to the unassisted body.

The third value of immediacy is of those things which are *perceivable.* These are things which can interact in an informational way with a human being (or, allowing at most one, nonartifactual intermediary). This category is meant to cover those things that one can see, hear, feel, etc., without the assitance of observing tools. But even something so simply observable as the moon does not itself interact with the human body, (well, it does gravitationally, but we are talking about seeking the moon) and hence the allowance for one intermediate interaction, a direct informational messenger. In the case of the moon, the medium is the group of photons which first intereact with the moon and then with the pigment in the eye.

These three categories are meant to reflect an intuitive implementation of the interaction-information picture of observability. The unobservable in principle, unperceivable in fact, and perceivable descriptions represent, respectively, cases where no informational interaction is possible, where informational interaction is possible but only through observing tools, and where informational interaction is possible without tools. I will use this trichotomy in the following survey of examples and analyze examples in each of the three categories. I do this both because it is a helpful organizational device, and to see what relevance and what amount of scientific appropriateness this division has. But, the dimension of immediacy is just a step in analyzing the acquisition of information, an effort in *triage* to initially sort cases in terms of how best to treat them. The three values of immedicay are by no means detailed enough to separate importantly different cases. To add detail we must look at other dimensions, particularly of the middle class, the unperceivable in fact.

A second dimension which reveals the fine structure within the class of the unperceiveable in fact is that of *directness*. This dimension is a modification of Shapere's direct/indirect dichotomy and it asks of the number of intermediate messengers that are required to convey the information from the initial object-state to the final observing-state, and the complexity of their physical journey. Cast in thoroughly information theoretic terms, the measure of the directness of observability is the measure of the number of interactions which would have to take place along the interaction chain from the initial-state to the final-state in order for the information to be conveyed. Simply put, how long must be the interaction chain from the object to the observer?

For example, information acquired with the aid of a magnifying glass is somewhat less direct than information obtained with the unaided eye, and information processed through a compound microscope is even less direct. This is determined by noting, in the first case, the extra interaction between photons and lens, and in the third case, the many interactions between photons and lenses. Each of these interactions is, in the physical science description of the process, a synapse in the conduit of information and its reliability must be accountable.

The measure of directness, the counting of interactions necessary to transfer the information, is an indication of both physical length and complexity of the path of information. We can imagine a case of informative interaction in which there is but one intermediate messenger, a photon, say, that interacts initially with the object and later with an observer, but whose path between the two is a complex course of bends, bumps and ricochets. This complexity in the path is recorded in the dimension of directness since each bend, bump or ricochet is an interaction, a change of energy and/or momentum of the photon. Directness measures not just changes of objects along the interaction chain but also changes of properties in a single object, since either kind of change marks a change of information content.

Counting interactions may seem an overwhelming task, but it is a relatively commonplace activity in the context of the physical sciences. A demonstration of just how it is done is the business of the next chapter and it is there that I will make good on the promise that directness is measureable. But for now we can take comfort in the knowledge that even in the most complicated of cases, for example the tedious escape of photons from the interior of the sun, physics is a ready accountant of interactions. The procedure is to calculate mean-free-paths, a parameter reciprocal to directness.

A third dimension of observability is the *amount of interpretation* which is necessary to correlate the A-state information to S-state information. This is intended to evaluate the nesting of S-state information in the final A-state information. Given the A-state information (z is R), the amount of interpretation is a measure of how deeply nested is the S-state information (x is P). The measure can be quantified simply in terms of how many distinct physical laws are needed to get from (z is R) to (x is P).

The typical situation is this: We are presented with a final apparatus state with informational content that z is R. From the fact that z is R we are to infer that x is P, and this will require a certain number of steps and a certain number of laws of nature to license the steps. (These laws are of the form, 'all z that are R are also S', 'if z is S then x is P', and so on). It is the number of such laws, the amount of assistance from auxiliary theories, that is the measure of amount of interpretation. The image in an optical microscope, for example, bears information of the specimen, and our scientific account of the specimen-information must trace its passage by invoking a variety of theories. The chemistry of the specimen describes its differential absorbtion of stain. Physical optics accounts for the interaction of stain with light waves and, perhaps by interference (see Hacking 1981, Ch. 11), imparting information on shape of the speciment to the light waves. Geometric optics describes the reshaping (but not equivocation) of the information by interaction of light with lenses. These are the auxiliary theories, and insofar as observation accounts differ as to how many such theories are required, the accounts differ in their amount of interpretation.

The dimension of amount of interpretation is different from the previous one, directness. Something which is observable via many interactions is not necessarily dependent on a commensurately many laws to account for the information. Take Shapere's example of observing the interior of the sun using photons. This observation is hugely indirect since there is a very large number of interactions (and a large number of messengers) required to pass the information along. But since the interactions are nearly all of the same nature, the transmission of information is accounted for by a single physical law, and such an observation measures relatively low in terms of amount of interpretation. Amount of interpretation is a measure of epistemic closeness between the A- and S-states. Directness, on the other hand, is a measure of the physical closeness between the two states.

A fourth dimension of observability can be identified, that of *independence of interpretation*. Given that there is a theory T of the object x in question (a theory which the observation of < x, P > is likely intended to test), one can ask whether the physical laws which account for the delivery of information in observation are independent of T. The answer is the measure of independence of interpretation. The manner and amount to which the theory T figures in the inference in the observation of < x, P > is closely related to the question of the theory dependence of observation as raised by Fodor and Shapere. It is a question of whether T can be isolated from the process of observing x, whether, in other words, only theories which are not theories of the object itself can suffice to support the observation. By "support the observation" is meant that a theory is used to show that the information (x is P) is nomically nested in the final, observation apparatus information (z is R).

The concept of independence is initially rather vague, and since it will play an important role in what is to follow, it is worth the effort of clarification. There will be two notions of independence at work in the following analysis. One of these is a relation between two theories, the relation simply of independence. Given two theories T_i and T_j (considered as collections of propositions), we will want to be clear on what it means for T_i to be independent of T_j. What allows us to say, for example, that the theory of evolution is independent of the theory of relativity?

The other notion of independence is independence of an account (or, in the language of the interaction-information account, independence of interpretation). This will be more complicated than simple independence between theories since it asks whether the warrant for a putative observation claim is independent of the theorectical description of the thing putatively observed. That is, a claim to have observed that x is P, or to have gathered as evidence that x is P should be more or less independent of any theoretical claims that x is P, if the evidence is to be useful in confirming the theory of x and P. We have, for example, a theory of gravity which states that all unsupported objects accelerate toward the earth at the same rate g. The evidence of a particular stone falling at a rate of g functions in confirming the theory because the warrant for the evidence is not from the theory itself (say, by instantiating the general law) but by independent accounting based on principles of perception and optics. This independence of accounting, for which we want a precise formulation, is a relation between the set of theories $\{T_i\}$ used in the account of the observation, and the theory of the object and property in question.

Given the extensive detail in the description of the dimension of independence of interpretation, I should repeat that a general, intuitive grasp of the concept should suffice in understanding the following examples and their epistemological signficance. The details are in place only to arbitrate disagreements in evaluations of independence.

I shall describe, first intuitively and then more technically, how to evaluate and measure each of these two notions of independence. It is the latter, the independece of an account which will be the primary concern later on. But this kind of independence incorporates the other, so both must be described. In either case, I believe that an understanding of independence at the intuitive level is sufficient to follow my claims about observability and to appreciate the examples. Those readers who prefer to only skim the technical paragraphs can do so without endangering their ability to understand what follows.

The concept of indpendence of two theories should be based on epistemic justification. Circumstances of discovery of the theories, for example that they were both discovered by the same person or motivated by the same evidence, should not influence the evaluation of independence in this sense. The two theories are independent in an epistemologically significant way if the truth or falsity of one is isolated from the truth or falsity of the other.

On an intuitive level then, say that T_i is independent of T_j if and only if T_j being false in any way would not indicate that T_i is false in any way, and T_i being false in any way would not indicate that T_j is false in any way. As an example, Newtonian mechanics is likely to come out as independent of the theory of evolution on this analysis since a realization that either of the theories is flawed would not be reason to question the other.

This characterization of independence can be made more precise if we regard a theory as a collection of propositions. Note that these theoretical propositions are ususally generalitzations (about *all* falling bodies and not about this particular stone). A theory T_i is false ($-T_i$) if at least one of its member propositions is false. With this we can define the relation of theory independence: T_i is independent of T_j if and only if it is not the case that there is a proposition p_i in T_i such that $-T_j$ nomically entails that p_i is false, and it is not the case that there is a proposition p_j in T_j such that $-T_i$ nomically entails that p_j is false.

This relation defines a dichotomy. Once T_i and T_j have been identified they either are or are not independent in this sense. There will be an element of caprice, however, in determining which propositions to include in the collection which is the theory. That is, if there is some proposition p_i in T such that $-T_j$ entails that p_i is false, one might simply jettison p_i from the theory and deal with a new editted version of T_i which is independent of T_j.

The second notion of independence, independence of an account, is neither a dichotomy nor subject to the arbitrariness of determining which propositions must be included to form a whole theory. The characterization of this brand of indpendence should indicate the degree to which a singular claim (that x is P) is supported by an account which is independent of the theory of x.

The first step in this characterization is to identify T_x1 as that sub-theory (sub-collection of propositions) of T_x, the theory-of-x, for which the observation (the singular claim that this x is P) could be used as confirmation. T_x1 will be a collection of propositions for which the fact that x is P could be used as confirming evidence. For example, the claim 'x is P' may function as a premise in an abductive argument for a proposition in T_x1. Clearly, T_x1 will be a collection of claims (generalizations) about x's and P. And it is this sub-theory T_x1 that one wants to isolate from the account of observing that the particular x is P.

The next step in the characterization is to identify T_x2 as that sub-theory of the theory-of-x T_x that is part of, or simply not independent of, the collection of theories $\{T_i\}$ used to support the claim of having observed that (having evidence that) x is P.

The motiviation in identifying T_x1 and T_x2 is the realization that, for some theory-of-x T_x (examples: theory of electrons, theory of plate tectonics, theory of black holes) there may be parts T_x2 used to account for some evidence (example: how tectonics plates interact with seismic waves) and parts T_x1 that describe particular properties which are putatively observed (example: the motion of a tectonic plate). The measure of the independence of account will be an evaluation of both the relation between T_x1 and T_x2, *and* the relation between T_x2 and $\{T_i\}$. A high score for independence results in the case of little or no intersection of T_x1 and T_x2, and, in the same case, T_x2 playing only a minor role in the account $\{T_i\}$. The role of T_x2 is minor if it is not actually part of the $\{T_i\}$ but is only not-independent, or if Tx2 is only a small part in $\{T_i\}$, assuming that it is an epistemic virtue to demonstrate an ability to cooperate with a large field of theories.

The point in this intuitive explication of independence of an account is that part of the theory-of-x may be used to account for the observation that this x is P, but a degree of independence is preserved if it is not a crucial part of the theory, the part that explicitly discusses x and P and could therefore benefit (by confirmation) if in fact this x was P.

This characterization needs a more precise description of "the relation between T_x1 and T_x2" and of "the relation between T_x2 and $\{T_i\}$". Different possible states of these relations result in a scale of degrees of independence of an account. At the top of the scale, the most independent, is the case where T_{x2} is empty, that is, no part of T_x is used in the account.

Of the cases in which T_x2 is not empty, the best, from the point of view of independence are those in which T_x2 and T_x1 are disjoint. (This list of the degrees of independence is summarized below in Table 1). Within this case, it is best if T_x2 is not a member of $\{T_i\}$ (but is only not-independent of at least one T_i). Then, still in the case where T_x2 and T_x1 are disjoint, descending toward the less independent is the situation of $\{T_x2\}$ being a proper subset of $\{T_i\}$. In this case, T_x2 is acting in support of the account, but can do so only in cooperation with other theories. Last, in this case of disjoint T_x2 and T_x1 is the case of $\{T_x2\} = \{T_i\}$. In this case, T_x2, T_x1's theoretical crony takes full epistemic responsibilty for the account.

There is another, less independent case of the relation between T_x2 and T_x1, that is if part, but not all of T_x1 intersects T_x2. This is clearly heading into the realm of not-so-independent since part of the theory that the observation could confirm (or be derived from) is used in, or is at least not independent of, the account of the observation. This case of intersection between T_x1 and T_x2 can be subdivided into the same three cases of relation between T_x2 and $\{T_i\}$ as the previous case. This time though, it is not just theoretical cronies of x and P that are not-independent of, part of, or all of the accounting of the observation, it is members of the immediate theoretical family. We've slipped from theoretical cronyism to a kind of theoretical nepotism.

Continuing down the path toward non-independence, there is the case in which T_x1 is a proper subset of T_x2. This indicates that our account of the observation that x is P is such that no theoretical claim about x and P is indpendent of the theories used to justify the observation claim. And since the same three cases of relation between T_x2 and $\{T_i\}$ apply here as above, there is a worst case within this case in which *all* theoretical claims about x and P are used in the account.

But things could be even worse, or at least less independent. The least independent cases result if T_x2 is in fact nothing but T_x1 itself. And given the now recycled three degrees of relation between T_x2 and $\{T_i\}$ we come to the least possible case of independence, that of $T_x1 = T_x2$ and $\{T_x2\} = \{T_i\}$. In this case our support for the putative observation that x is P is based entirely on the whole of the theory about x being P.

This completes the spectrum of independence of an account. These results are summarized on Table 1 which lists the cases in descending order of degree of independence.

Closeness of relation between Tx1 and Tx2	**Depth of involvement of Tx2 in {Ti}**
$T_x2 = 0$ most independent	
$(T_x2 \;\; T_x1) = 0$ theoretical cronies	T_x2 in $\{T_i\}$ T_x2 in $\{T_i\}$ but not $\{T_x2\} = \{T_i\}$ $\{T_x2\} = \{T_i\}$
$(T_x2 \;\; T_x1) = 0$ but not $T_x1 \;\; T_x2$ theoretical nepotism	T_x2 in $\{T_i\}$ T_x2 in $\{T_i\}$ but not $\{T_x2\} = \{T_i\}$ $\{T_x2\} = \{T_i\}$
$T_x1 \;\; T_x2$ but not $T_x1 = T_x2$	T_x2 in $\{T_i\}$ T_x2 in $\{T_i\}$ but not $\{T_x2\} = \{T_i\}$ $\{T_x2\} = \{T_i\}$
$T_x1 = T_x2$ least independent	T_x2 in $\{T_i\}$ T_x2 in $\{T_i\}$ but not $\{T_x2\} = \{T_i\}$ $\{T_x2\} = \{T_i\}$

Table 1: Degrees of Independence

One can see a priori that there is a difference between the amount of intrepretation and the independence or dependence of interpretation. The informational account of an observation could involve extensive inference, the information could be deeply nested, yet all of the inference could be independent in the sense that none of the laws accounting for the nomic nesting are T-laws. Or an observational process could involve only a single inference which involves T. The point is that there is no obvious connection between amount and independence of inference. The dimension of independence of inference does not measure epistemic closeness, but epistemic nepotism.

The reason for pointing out all these dimensions of observability (and there may even be more) is to demonstrate the complexity of the concept. Classification of observability is neither a dichotomy between the observable and the unobservable, nor even a continuous spectrum along one axis. It is an evaluation of several parameters, locating the object-property pair in a many dimensional observability space. Evaluating each parameter requires consulting the physical science descriptions of the object x, the interaction-information chain, and the observing apparatus A-state. This is a capitulation to the theory dependence of observability, though not at all to the theory dependence of observation as this latter issue is clarified above.

The realization of Shapere's, Achinstein's, van Fraassen's and Fodor's suggestion to use physical theory to decide questions of observability, that is, to regard observability in science as an empirical question, is to invoke physical theories to describe intereactions and to follow the trail of information. This latter task is accomplished by uncovering the nomically nested information.

In deciding questions of observability and the reliability of information-acquistion reports, we are always confined to the final (apparatus) informational states and the current scientific account of the chain of interaction and flow of information. We cannot indulge in a perspective outside of the scientific endeavor from which all the facts are known and the primary task is to decide what counts as observable. Rather, this study must be done from a perspective internal to science if it is in any way to assist science in deciding what the world is like. The dimensions of observability described in this chapter are all well suited to this internal perspective and are, in the end, intended to be useful in assessing the reliability of claims about what there is in the world.

The general guidelines developed in this chapter for studying observability tell us at least this much. The question of whether or not an object or even an object-property pair is observable is a bad question. It is a bad question because it ignores the complexity of observability. It looks for a one dimensional, discrete observability space where, in fact, that space is many dimensional and likely continuous. The better question is one very similar to that advocated by Achinstein. Ask instead about the observability characteristic of the object-property pair. Specifically ask, what is the nature of the information of the object-property pair and how does it get from the object to the observing apparatus. Then answer these questions by locating the example in observability space, that is by evaluating its components along the dimensions of observability.

CHAPTER 3

CASE STUDIES

1. THINGS UNOBSERVABLE IN PRINCIPLE

Begin the survey of examples of observability with those entities which, according to the relevant theories, cannot interact with any kind of observing apparatus in a way which conveys information of the entity.

Things causally isolated in spacetime

That there are things which are, because of their spacetime location, causally isolated from the observer, is a consequence of the finite propagation speed of causal influences. Hence, these things are of no concern in spacetimes such as the Newtonian spacetime with possible infinite causal speeds. But any physical theory which is consistent with special relativity will include a finite upper bound to the speed of information and will have to reckon with the status of things which are unobservable because of this restriction. These will be things which occupy a spatial location such that in principle no interaction effect could reach the observer by some specified time t. Different restrictions on what is a relevant time t, whether it is t = now, or t = the death of the observer, etc., will present different cases of this kind of unobservability, some of which will be more important than others.

It may seem that something which is unobservable because of its spacetime location is unobservable in fact but not in principle. Admittedly, it is a contingent property whether a particular object is at a spacetime location to be causally isolated from an earthbound observer and hence unobservable. But for the general class of objects in these circumstances it is a matter of principle that they are unobservable in the sense that there is nothing physically possible (as allowed by the spacetime theories being considered) that can be done to bring about interaction between the object and an accessible apparatus. The object cannot be moved into an observable location, nor can the observer move within range. Bright lights, telescopes, shrinking the apparatus and enlarging the object all have no effect. Given that something is at such an unobservable point in spacetime, no interaction is allowed. But since there is no particular kind of thing which could be in this situation (it could be a star, a quark, or my bicycle; anything at all), it is best to regard this causal-location status of observability as a general condition of observability status. That is, one should think of a causally connected location as a general prerequisite for anything to be observable. Nonetheless, the status of being unobservable because of spacetime location is an interesting condition which needs to be understood in its various degrees.

Observability is best described in terms of an object-property pair, and in this case there are at least two kinds of "objects" which present interesting cases. One can focus on an individual event and ask whether its spacetime location precludes the availability of information about such a basic property as the fact that it occurred. Or, consider an enduring object with an extended worldline, call it a particle, to see if the entire length of the worldline is unobservable because of its spacetime location. In this case, information on the properties on the object such as its mass, angular momentum and even its presence, will be kept from the observer. The particle itself may be unobservable throughout its existence. The two different kinds of "objects", events and particles, or, in spacetime terms, points and worldlines, require different circumstances for principled unobservability. Principled unobservability because of spacetime location can be the result of an event horizon or a particle horizon. In either case, information about the basic properties of the event or particle cannot reach the observer owing to the relative situations in spacetime and causal structure of spacetime.

We will be interested in cosmological spacetime models which are consistent with our current empirical understanding of the global properties of the universe. In this interest, one can specialize to spacetimes which are spatially homogeneous and isotropic, since these properties seem to be supported by observation. It should be pointed out that this is not a great restriction since most viable cosmological models are homogeneous and isotropic. In this context that are four distinct cases of principled unobservability.

In the interest of limiting the topic to a manageable size I plan to ignore cases of local curvature singularities where particles and events are unobservable within event horizons such as surround black holes. This kind of unobservability might be more interesting if it did *not* happen than if it did. That is, if there was a singularity that was not hidden within a horizon, a naked singularity, then the crazy violations of physical theories would be exposed to view. The conjecture that these crimes against nature are always unobservable, that singularities formed in gravitational collapse are always surrounded by horizons, is the Cosmic Censorship hypothesis (first suggested by Penrose, 1969). This hypothesis is often considered as "the most important unsolved problem in classical general relativity" (Tipler 1985, 499), and it is clearly an interesting issue of observability. For the sake of brevity though I will not deal with spacetime singularities other than the possible initial big bang (which is a naked, uncensored singularity (Hawking 1976, 2461)), and a final collapse. The kind of singularities whose nakedness might be censored are precluded with the stipulation that the spacetime models being considered are homogeneous.

It is most convenient to describe the general homogeneous, isotropic spacetime model in terms of comoving coordinates. This is a coordinate system which is fixed to the large scale objects of the universe such as galaxies, such that the coordinate locations of such objects and the coordinate intervals between any two are always the same. Any change, homogeneous expansion or contraction, is accounted for in the spacetime metric. The general homogeneous, isotropic spacetime metric using these comoving coordinates is the Robertson-Walker metric:

$$d\tau^2 = dt^2 - R^2(t)\left\{\frac{dr^2}{1-kr^2} + r^2\, d\theta^2 + r^2 \sin^2\theta\, d\phi^2\right\}$$

(For convenience, the speed of light has been set equal to one). $R(t)$, as a function of coordinate time, accounts for expansion or contraction of physical three-space. The constant k can be one of three values, $k = -1, 0, 1$, corresponding to models of space which have negative, zero, or positive curvature, respectively. Minkowski spacetime is a special case of the R-W metric, and so local objects and events in the laboratory where spacetime is Minkowskian, can be included in this treatment.

To describe the conditions of unobservability it is most convenient to put the observer at coordinate location $r = 0$. Since the spacetime is homogeneous this gives the observer no privileged position whatsoever. We can also simplify the description by ignoring the angular coordinates. This move is sanctioned by the property of isotropy of the spacetime. With these simplifications it can be said that an event located at coordinates r_1, t_1 can interact with the observer at time t, only if r_1 satisfies the equation

$$\int_{t_1}^{t} \frac{dt'}{R(t')} = \int_{0}^{r_1} \frac{dr}{\sqrt{1-kr^2}}$$

This is derived from considering the null geodesics which represent the fastest interaction signals which can propagate from the event at (t_1, r_1) to the observer at (t, 0). The equation prescribes the relation between the observer such that anything located further away than r_1 (at t_1) will be unobservable in principle at time t. There are four different cases to be considered. The cases are distinguished by the difference in coordinate time t_1 of the event and the relevant coordinate time t of observation.

Case 1: things at $t_1 > t_{min}$ and observations at $t < t_{max}$

For studying observability as a function of spacetime location it is important to consider the extremes of coordinate age of the universe. That is, one wants to speak of worldlines which extend back in time as far as possible, worldlines that have existed since the beginning. The time at which such worldlines begin can be referred to as t_{min}, and it is either infinitely long ago ($t_{min} \to -\infty$) for models with no initial singularity, or it can be set to zero ($t_{min} = 0$) for models with an initial singularity, a big bang. In either case, regard t_{min} as the earliest possible time. Similarly, regard t_{max} as the most distant future time. So the time toward which eternal world lines are headed is either infinitely later ($t_{max} \to \infty$) for models with no final singularity, or it will be some finite value which is the time of collapse, for models with a final singularity.

Consider as a first case events and observations which happen strictly within but not on the boundaries of time. That is, consider events which occur at $t_1 > t_{min}$ and observations which are done at $t < t_{max}$. Whatever such time t_1 that is chosen, there will be spatial locations r_1 such that an event at (t_1, r_1) will be unobservable in principle at time t, simply because the event is too far away for a causal signal to reach the observer by time t. Simply put, this is saying that there are always events outside an observer's past lightcone at t. This is true of all the spacetime models being considered. It is a classification of unobservability which is a function of the time t, and as such it loses importance by being as arbitrary as one's selection of the relevant time t. If one chooses t = now, then the things under consideration are unobservable-up-to-the-present. This group includes not only recent (as measured in coordinate time) events at the edge of the universe, but also such events as a solar flare which happened just one minute ago or tomorrow's sunrise. These are events which are unobservable now but which will become observable in the future at the point where interaction signals can reach the observer's worldline.

One might extend this class of unobservables by setting t = the time of death of the observer, or even less personal, t = the time of death of the last human. Then events outside the past light cone at this time will be unobservable-within-a-lifetime. This still carries a certain arbitrariness with respect to when the observer or the species might die, a sort of event which is subject to accident or foolishness. It does not seem to be a useful or meaningful classification of observability which is so sensitive.

Whatever time t one chooses as the cutoff of observability, this kind of unobservability refers to events but not to enduring objects. To say that certain solar events are unobservable now or within a lifetime is not to say that the object, the sun, is unobservable. To evaluate the observability of an object one must consider the entire worldline of that object, and in the case of the sun there are parts of the worldline which are within an earth-bound observer's past light cone. This situation is clearly shown in a spacetime diagram. (figure 1)

This and other spacetime diagrams which follow are generally representative of any R-W metric. The light cone can always be drawn as straight lines on the page by using conformal transformation of the spatial coordinates.

This notion of unobservability can be extended to include enduring particles simply by considering all events of the particle's worldline. An object is unobservable in this sense if no point of its worldline lies in the observer's past light cone at t.

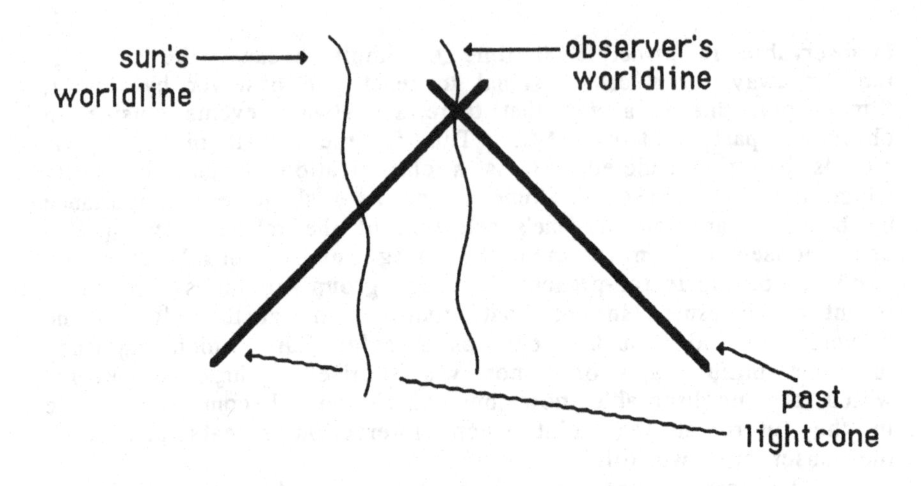

figure 1

The essential aspects of the case of events which happen strictly within the bounds of time and are unobservable because of being outside the observer's past light cone can be summarized. This sort of unobservability is ubiquitous to the point of being uninteresting. All spacetimes (with a restriction on the propagation speed of information) include such events. This classification is also dependent of the choice of relevant time t in a way that seems undesirable. Whether t is now, or the moment the research paper is published, or the death of the species, this does not seem like something that out to figure into a meaningful classification of observability. The more interesting cases of unobservability will be those in which the spacetime itself chooses t, that is where $t = t_{max}$. It is also interesting to ask whether there can be objects which have been around since the beginning (t_{min}) whose worldlines lie entirely outside an observer's past light cone and are thereby unobservable at present. To answer this, consider the next case.

Case 2: things $t_1 = t_{min}$, and observations at $t < t_{max}$

This case is asking after objects such as galaxies or stars which have existed forever (or since the initial big bang) but which are unobservable at some t (now or in a lifetime). It also considers regions of space which may or may not be occupied with stuff, which are unobservable in principle (if it even makes sense to evaluate a region of space in term of observability). To see if a particular spacetime includes such unobservables one must evaluate the integral relation

$$\int_{t_{min}}^{t} \frac{dt'}{R(t')} = \int_{0}^{r_1} \frac{dr}{\sqrt{1-kr^2}}$$

If the time-integral converges, then there is some value of r_1, which satisfies the equation, and anything at a spatial location greater than r_1 will be unobservable at time t. Such an object is said to be outside of the *particle horizon* of the observer at time t. The particle horizon is determined by the past light cone of the observer at time t, extended as far back in time as possible. It represents the boundary between the observable and the unobservable. Like the first case considered, this unobservability status is a function of time. But this case is more interesting that the previous one since it looks as far back as time goes and it speaks of enduring objects that have existed forever but are unobservable at t. Presumably, the observations of astronomy which have been used to formulate current astronomical theories cannot have accounted for such things.

Not all spacetimes have a particle horizon. For Minkowski spacetime which is flat ($k = 0$), static ($R(t) = 1$), and temporally unbounded ($t_{min} \rightarrow -\infty$), it is easy to show that the time-integral diverges for any value of t. So in Minkowski spacetime there are no unobservable worldlines that begin at t_{min}. That is, for enduring objects that have existed since the beginning, there is no place in space for them to hide in causal isolation. One can observe all of space and all its enduring contents.

Any Robertson-Walker model which is positively curved ($k = 1$), and even some that are flat (for example, the deSitter metric: $k = 0$, $R(t) = \cosh^2 t$, $t_{min} \rightarrow -\infty$) will have a particle horizon. And since these are the models most often used to describe the physical universe, it is accurate to say that most cosmological models include the possibility of objects which are unobservable because they are beyond the particle horizon. But all of these models require that there exist some massive objects beyond the particle horizon, since the spacetime is assumed to be spatially homogeneous. To say that a spacetime is spatially homogeneous is to say that all areas are metrically equivalent. In the large scale such that effects of single stars and galaxies are averaged out, no area of space is distinguished by its curvature. With the general-relativistic connection between curvature and matter, this means that no area of an homogeneous space is distinguished by its mass content. The stars and galaxies are spread uniformly through space. This is known as the "Cosmological Principle" (Weinberg 1982, 407). It is motivated by observation of that part of the universe which is observable, and it is the basis for using cosmological models which are homogeneous. These models then describe all space, observable or not, as being equally densely filled with material objects. In this sense the, the consistency of these cosmological models requires existent, unobservable objects.

To get an idea of what fraction of the population of astronomical objects are outside of the particle horizon it is necessary to consult the empirical evidence which is used to estimate the relevant cosmological parameters such as the size of the universe ($R(t_0)$, where t_0 = now), the deceleration parameter, the Hubble constant, etc., Weinberg (1982, 940) summarizes the data and calculations and concludes that for any positively curved, dynamic model which expands to a maximum size R_{max} and collapses back to a singularity $R = 0$, half of physical space will fall inside the particle horizon at the time of largest extent. For such a model with physical parameters that match measured values, at the present time the particle horizon extends roughly to only 1/4 of the size of the universe.

The significance of a particle horizon is best appreciated with a spacetime diagram On a diagram where the coordinates are conformally transformed such that past and future null infinities can be mapped onto a finite picture, it is apparent that a particle horizon for a nonaccelerating observer results whenever the past null infinity $\mathscr{I}^-$ is spacelike (Hawking and Ellis 1973, 129). ($\mathscr{I}^-$ is the region from which null geodesics which begin at infinity come). There will be not particle horizon if $\mathscr{I}^-$ is null.

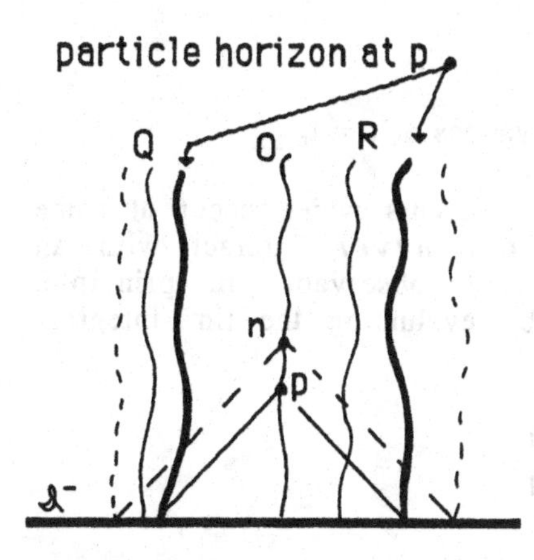

(a) spacetime with a particle horizon

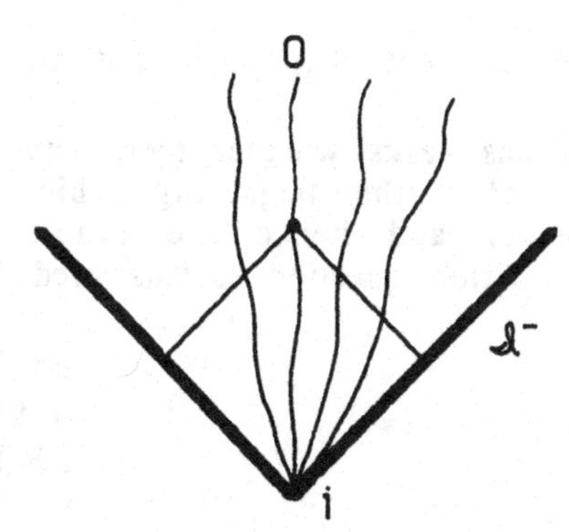

(b) spacetime (eg.Minkowski) with no particle horizon

figure 2

Figure 2a represents the particle horizon at point p of an observer O. By the observability criterion of the particle horizon, R is an observable particle but Q is unobservable in principle at time t (point p). It is possible however that Q will become observable at some later time, that is, that there will be some later time (point n) on the worldline of O such that the worldline of Q is within the particle horizon at n. Figure 2b shows a spacetime such as Minkowski spacetime in which the surface of origin of null geodesics is null and all time-like geodesics, that is all worldlines of eternal particles, originate from the point i^-. In this case all particle worldlines have points within the past light cone of observer O at all times, and hence there are no objects which have exited forever but are unobservable in principle because of their spatial location.

It is important to remember that though this unobservability status does not depend on a privileged spatial location of the observer, it does depend on the particular choice of time t at which the particle horizon is calculated. To get an unobservability condition which is not as arbitrary as one's choice of t, one can use the one time value distinguished by the cosmological model itself, namely t_{max}. This leads to two more cases of unobservability.

Case 3: events at $t_1 > t_{min}$ and observations at $t = t_{max}$

This case asks whether there can be events which occur at some time after the beginning which can *never* interact with an observer and hence are timelessly unobservable in principle. Again, this question is answered by evaluating the time-integral

$$\int_{t_1}^{t_{max}} \frac{dt}{R(t)}$$

where t_{max} is the time of the future singularity for models that predict collapse, or $t_{max} \rightarrow \infty$ if there is no future singularity. If this integral converges, there will be spacetime locations such that events at these locations can never interact with the observer. The boundary which divides events which will be unobservable forever from those which will be observable sometime is the *event horizon*. It is simply the past lightcone of the observer at the time t_{max}. As with the particle horizon it is most clearly understandable with a conformal spacetime diagram.

Figure 3a shows that for a nonaccelerating observer a spacelike future null infinity $\mathscr{I}^+$ (the region to which null geodesics which end at t_{max} go) is the sufficient condition for there being an event horizon (Hawking and Ellis, 1973, 129). The event at point s, for example, is unobservable to the observer in the sense that it can never interact with O. This is an unobservability in principle which does not require any specification of the time of observation. But this classification is best restricted to events rather than enduring particles. This is because it is possible to have some events on an object's worldline be outside the event horizon, but have some segment of the worldline inside the horizon. In the figure 3a, the object Q is unobservable in this sense only for times after point r. The object Q itself, one would want to say, is observable since at least part of its history is situated such that it can interact with O. Here then it is important to be clear on what sort of object one inserts into the object-property pair of which to evaluate observability. An event, a point in spacetime, figures relevantly in this case, but an enduring particle, a timelike line extending from i^- to i^+, does not. The point is not that one kind of thing, event or particle, is the more appropriate subject of observability, but only that care should be taken to distinguish events from particles in making observability claims.

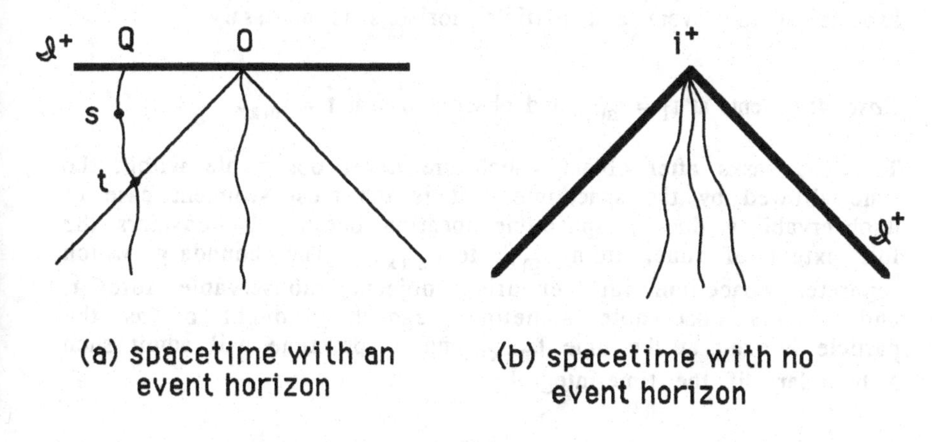

(a) spacetime with an event horizon

(b) spacetime with no event horizon

figure 3
(based on Hawking and Ellis 1973, 130)

Figure 3b shows that null future null infinity $\mathscr{I}^+$ disallows there being an event horizon. All events are such that they will be able to interact sometime.

Again it is interesting to evaluate the viable cosmological models in terms of their having event horizons, using the empirical evidence at hand. The deSitter spacetime which is the model used for many steady-state cosmologies does have an event horizon. And Weinberg (1972, 490) calculates that any cosmological model with a deceleration parameter $q_0 > 1/2$ will have an event horizon. Taking $q_0 = 1$ as a reasonable compromise among measurements, Weinberg calculates that of the events occurring now (t_1 = now) only those within a proper distance of 61×10^9 light years will be within the event horizon. The present extent of the universe is estimated to be 80×10^9 light years. Events happening now but further away then 61×10^9 light years can never be observed.

This classification of unobservability is independent of any choice of time observation, but it is still a function of the time t_1 of the event. This makes the criterion inappropriate for deciding whether an enduring object will be unobservable. To extend this category to objects, a final case which combines aspects of the event and particle horizons is necessary.

Case 4: events at $t_1 = t_{min}$ and observations at $t = t_{max}$

This case asks after object which are never observable within the time allowed by the spacetime. It is the most stringent case of unobservability due to spacetime location because it considers the full extent of time, from t_{min} to t_{max}. The boundary which separates spacetime into enduring objects unobservable forever, and objects observable sometime, can be thought of as the particle horizon at the time t_{max}, and a spacetime will admit such a boundary if the time-integral

$$\int_{t_{min}}^{t_{max}} \frac{dt}{R(t)}$$

converges. Convergence indicates that these are objects which exist from start to finish of the universe (that is, forever) but which are unobservable in principle because any causal influence they send toward the observer cannot arrive before t_{max}. That is, there are regions of physical space that one has not only not been able to see up to now, but which one never can see even if one waits forever or until the universe collapses to a singularity.

The spacetime diagram of this situation includes both the past and future null infinities, and they must both be spacelike. (figure 4)

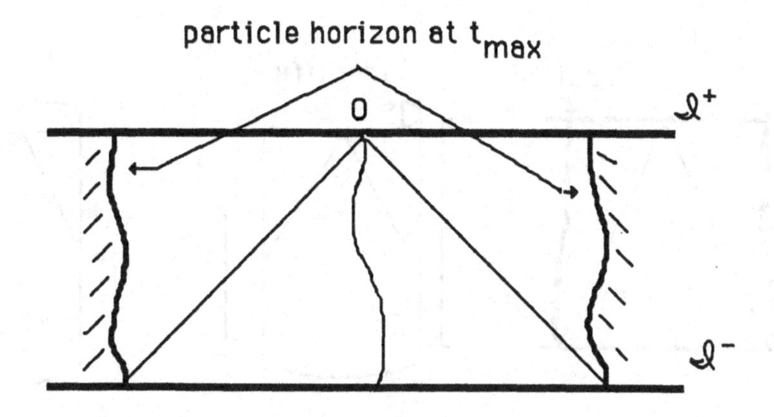

figure 4

An interesting consequence for spacetime models which have this feature of including regions of space which are absolutely unobservable is that there may also be aspects of global topology that are unobservable. This is a point made by Glymour (1977) and Malament (1977). As a specific example (which I take from Malament), consider the special case of a Robertson-Walker spacetime M such that $d\tau^2 = dt^2 - \cosh^2 t\, dr^2$. It is easy to evaluate the time-integral for $t_{min} \to -\infty$ and $t_{max} \to \infty$ to show that this model has a particle horizon at t_{max}. That is, it includes regions of space which are eternally unobservable. Now construct a new spacetime M' by cutting along the t_{max} particle horizon, removing the unobservable regions of space, and rolling the spatial dimension of M into a closed topology by identifying points of equal time-component on the cuts (figure 5).

This new spacetime M', which is just the DeSitter spacetime, has a topologically closed three-space, but it is observationally indistinguishable from M which has an open three-space. The important point is that not only can objects and events be unobservable in principle, but so can the global spacetime topology.

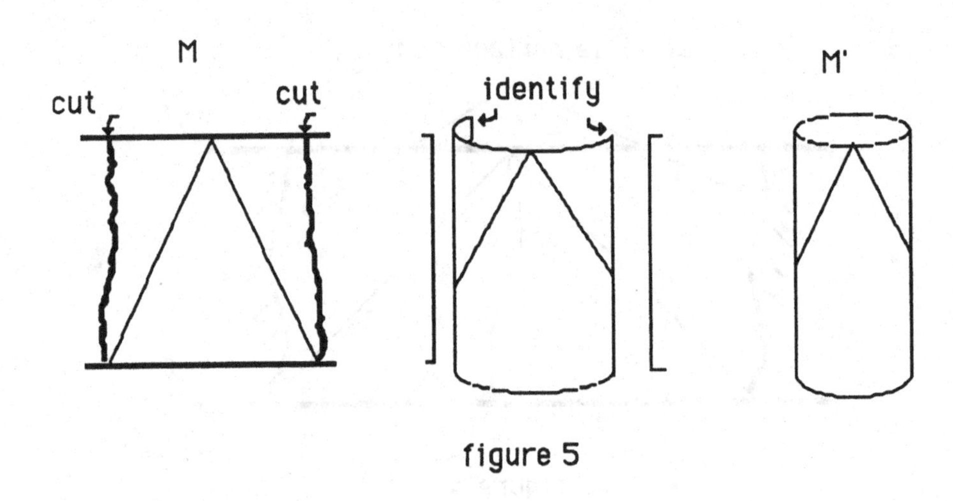

figure 5

In terms of the object-property pair description, say that some spacetimes (a funny kind of object, but a bearer of properties nonetheless) are in principle unobservable with respect to certain topological properties such as spatial closure. This unobservability is attributed to the fact that there is not enough time, even if there is infinite time, for a signal bearing the topological information to interact with the observer. There is nothing for it except to invoke Hawking's "principle of ignorance" (1976, 2462) which is to say that all data hidden behind an event horizon compatible with observable data are equally probable. In the case of the spacetime with metric $d\tau^2 = dt^2 - \cosh^2 t \, dr^2$ there is no empirical basis to favor one of M or M' over the other.

To summarize this survey of unobservability due to spacetime location, note first that once it has been decided that something is unobservable in principle there is little left to say about the other dimensions of observability. Since the possibility of a chain of interaction between object and observer has been precluded, there is nothing of which to measure directness. And with no object-information reaching the observer, there is no amount of interpretation which will unnest information of the object. Our policy of *triage* marks these cases as being beyond help, there being no chance of finding life or information in the patient.

There is something to be said though about the independence between the theory of the object and the theoretical account of its observability. In one sense there is a great deal of independence since the spacetime theory, the general theory of relativity and a particular solution as a model, has nothing to do with the detailed features of the astronomical objects in question. Theoretical descriptions of the properties and activities of stars and galaxies, the kind of information one seeks in astronomical observations, are not intimately part of the theory of spacetime. On the other hand, there is the dependence between theory of observability and theory of object that the cosmological model which prescribes unobservability in principle also prescribes existence of these things. Consistency with the assumption of homogeneity demands a nonempty set of unobservable massive objects. And it is important to note that the principle of unobservability in this sense arises naturally from the theory. It is not a case of postulating objects which are unobserved (such as individual quarks) or conceptually troublesome (the naked singularities mentioned earlier) and then rendering these things unobservable by fiat (color confinement and cosmic censorship, respectively). In the cosmological case the principle of unobservability is a consequence of a consistent theory.

A general conclusion to draw from the example of unobservability owing to spacetime location is that there is not a single circumstance or reason which accounts for such unobservability. Even within this subclassification of unobservability, there is an important variety of degrees among examples. The four cases were presented in increasing order of stringency, and not all cases of unobservability due to spacetime location are of equal epistemological or ontological importance. The first case of unobservables, which included tomorrow's sunrise, clearly offers a less serious criterion on which to base one's ontology than the division of a particle or event horizon.

Furthermore, different kinds of objects of observation figure differently in the observability classification. Evaluating the observability of events, particles, or topological features requires different standards as is shown by the difference between particle and event horizons. Hence, even though we have not been able to evaluate the dimensions of observability in this case we have learned something about the diversity of unobservables.

Single quarks

In switching attention from theories of spacetime to theories of elementary particles one changes the focus of inquiry from things in general to a specific kind of object. That is, all objects, large or small, simple or complex, are treated equally with regard to spacetime observability. That kind of principled unobservability is independent of the kind of object. On the other hand, an observability status which is decided by elementary particle physics will be sensitive to the differences in particular kinds of particles and perhaps even to particular circumstances in which the particle is located. There is at least one elementary particle which the current theory rules to be unobservable in principle, and that is the isolated quark. As part of the study of principled unobservability, it is worth understanding just what it means to say that an individual quark is unobservable.

Subatomic particles (other than photons) are divided into two disjoint groups, leptons, like electrons and neutrinos which do not participate in the strong nuclear interaction, and hadrons, like protons and pions which do interact via the strong nuclear force. There are only six known leptons, but over a hundred known hadrons. But whereas the leptons are regarded as genuinely elementary particles without any simpler constituents, the hadrons are now thought of as being compounds of quarks, where the quarks themselves are elementary. All baryons (a subset of hadrons) are composed of three quarks, and all mesons (the other subset of the hadrons) are composed of two quarks.

The quarks must exist in enough different varieties to account for the number of distinct combinations of threes and twos to make up the different baryons and mesons. Originally, the different kinds of quarks were distinguished by a single quantum number, their isospin. Different values of isospin correspond to different particles, that is, to different quarks. Isospin is also referred to as flavor, and one says that quarks come in different flavors. At last report, six flavors, that is, six isospin values, were needed to make up the variety of hadrons. The different values have been given very unflavorlike names: There are up, down, strange, charmed, top and bottom quarks.

To account for some of the hadrons as quark combinations, and to fit the description of quarks into quantum mechanics, it is necessary that quarks have another degree of freedom besides their flavor. Quarks must also have color. The reason that the color quantum number is necessary is best explained with an example. Because of the different fractional charges of the quarks, there is only one way to make the baryon Δ^{++} out of three quarks, and that is with three up-quarks with their spins aligned and all three in the ground state. But since all quarks are spin-1/2 particles they are the subject of the Pauli exclusion principle which is one of the pillars of quantum mechanics. The exclusive principle states that no quantum system can exist with identical particles of half-integer spin that are in identical states. In other words, the Δ^{++} cannot exist with the identical up quarks whose spins are 1/2, in identical states. There must be some difference between the three quarks. For this reason, one postulates the quantum number of color and allows it to have three distinct values, red, green, and blue. So in the Δ^{++} the three up quarks are in different color states. There is one red up-quark, one green up-quark, and one blue up-quark. The important thing to realize from this example is that color and flavor are different quantum numbers and that each quark has both a flavor value and a color value. But there is a crucial difference between flavor and color. It is in principle possible to measure flavor, but not color.

Quarks always exist in pairs as mesons or in triples as baryons. The meson state always consists of a color-anticolor pair, such as red-antired. (In the quantum description there is always an antiparticle which mirrors the properties of a particle.) The meson then has no color since the red and antired cancel. A baryon state always consists of three quarks which have different color values. So every baryon has one red, one green, and one blue quark as components. When the three possible quark colors are equally represented like this they sum to a colorless white, just as the colors of light which, when summed together, are white. So baryons, like mesons, are colorless. This property that quark color cannot be seen, either as showing through in a quark group or as from an isolated quark, is described as *color confinement*, and there is a clear theoretical reason why it must be imposed.

The argument for color confinement can be put as a reductio ad absurdum. Consider a single, color-charged quark isolated in vacuum. Just as an electrically charged electron disturbs the vacuum and incites the creation of a cloud of virtual electron-positron pairs and virtual photons, the quark disturbs The vacuum and is surrounded by virtual quark-antiquark pairs and virtual gluons. The difference between the quantum electrodynamic description of the electron and the quantum chromodynamic description of the quark is that in the former case the photon is electrically neutral but in the latter case the gluon is colored. The color-charged quark precipitates the creation of color-charged gluons, thus increasing the net color charge. This incites the creation of more color-charged gluons, further increasing the net color charge, and so on. An isolated color charge is unstable because its effect on the vacuum is to build up the charge without bound. This kind of instability cannot be allowed, hence isolated color charge cannot be allowed. Quarks can exist only in groups such that their color is neutralized.

To say that color cannot be seen is just a sloppy way to claim that a colored particle cannot interact with another object which we regard as a part of the observing apparatus in such a way that the apparatus is left in a different state according to whether the interaction was with a red, green, or blue particle. The color of quarks is unobservable because physical interactions are color blind. It is not a problem with our eyes or with our machines. The unobservability is dictated by the physical laws of interaction themselves. In terms of the interaction-information account of observability, the pair <quark, color> is unobservable in principle. No interaction between quark and observing apparatus can convey information of color.

Even though single quarks are unobservable with respect to color, there are some other properties of single quarks which can interact with the rest of the world. Such properties are then possibly only unperceivable in fact and not unobservable in principle. Quark flavor is a property that can interact in a correlated way with an apparatus. Consider, for example, charm. It is not that individual charmed quarks can exist in isolation, but rather that a charmed quark can exist in a bound state with an uncharmed quark (as opposed to an anticharmed quark) such as an up-quark. The resulting meson $D^0 = (c, \bar{u})$ is itself charmed since there is no anticharm to cancel the charmed quark. The charm of the quark shows through the meson in a way that color cannot. The charmed D^0 meson has a characteristic decay process (it is slow, hence the name "charmed") indicating that the world can tell whether it is charmed or not. Charm, in other words, has an effect on interactions in a way that allows the <quark, charm> pair to be called unperceivable only in fact but not unobservable in principle. For this reason, I think that Shrader-Frechette (1982) is, by describing charm, concentrating on the wrong quantum number to explain the unique unobservability status of quarks. It is not charm which makes quarks observationally distinct from other elementary particles like neutrinos; it is color.

It is important to understand what is meant when scientists talk about observing quarks, or as Fritzsch (1983) says using quotes, when quarks are "seen". The most distinctive quark events which are observed are quark jets. What is observed in a quark jet, that is, what actually interacts with the observing apparatus, are groups of hadrons which fly off from scattering events in relatively narrowly defined directions as show in figure 6a.

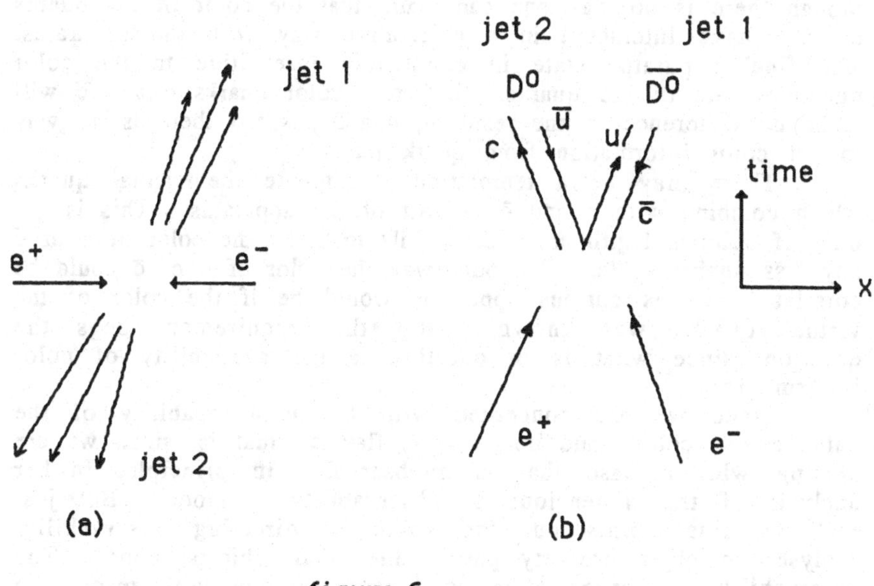

figure 6

These are called quark jets and are said to be indirect observations of quarks for the following reason. Theory predicts that the electron-positron collision produces a quark-antiquark pair (c, $\bar{c}$) which quickly splits, with the two constituents going off in opposite directions. But the separated c and $\bar{c}$ will each immediately find a mate in the cloud of virtual quarks which surrounds the disturbance. If each pairs with an up quark, the result is a D^o meson (c, $\bar{u}$) moving off in one direction and its antiparticle, a $\overline{D^o}$ meson ($\bar{c}$, u) moving in the opposite direction.

The D^0 and $\overline{D^0}$ decay into a shower of particles whose momenta are all in roughly the same direction. Thus the momentum of this ensuing jet of particles manifests the momentum of the original quark, and in this way the jets are informative of the quark process which initiated them. Thus are the quarks "visible". Figure 6b shows on a diagram with one spatial dimension running horizontally and time running vertically, the theoretical origins of the quark jets.

It is legitimate to regard the incident particles, the electron and the positron, plus the resulting hadrons in the jets as part of the apparatus of observation. Even with this extension though there is no way one can claim that the color of the quarks c or $\bar{c}$ have interacted in a correlative way with the apparatus. The final apparatus state in completely insensitive to the color quantum state of the quarks. Different color quarks c and $\bar{c}$ will make no difference in the resulting quark jets, so there is no way to get color information from quark jets.

There may be a temptation to include the virtual quarks which combine with c and $\bar{c}$ as part of the apparatus. This is not only of dubious legitimacy, but it fails to make the color of c or $\bar{c}$ any less unobservable. The only way the color of c or $\bar{c}$ could be correlated to this curious apparatus would be if the color of the virtual quarks was known. But this requirement begs the question, since what is in question is the availability of color information.

Since we are concerned with the unobservability of the pair <quark, color> and not <quark, flavor>, that is, since we are dealing with a case that is unobservable in principle, further analysis of the dimensions of observability is moot. But just realizing this points out the value of directing observability analyses to object-property pairs rather than objects alone. The observability properties of a quark are different with respect to color than to flavor. Furthermore, there is the warning in the analysis of quarks and color against untested generalizations about unobservables *per se*. It is evident that it would be mistaken to make a general claim about the unobservability of elementary particles in general. The reason for calling an electron unobservable is dramatically different from the reason for labeling a quark unobservable. While the former can interact with systems that convey information to an observer, there is an important sense in which the latter cannot. The epistemic status of an electron is importantly different from that of a quark.

Summary and conclusion

The purpose of presenting these two examples of unobservability has been to show that the concept of unobservability has importantly different meanings in different circumstances. There is an informative aspect to this study which is simply to point out how, in these two important cases, the appropriate scientific theories disallow observation in the sense of ruling out correlated interaction between the system in question and any apparatus of observation. It is valuable to know in each case just what kind of thing, whether it is an object, a property, or a state, that is being classified as unobservable, and just what it is about that thing which makes it so. This is an important concern of the issue of scientific realism since the outcome of that debate turns on the ontological status of the unobservable. To decide this, it seems prerequisite to understand just what it means to be unobservable. And insofar as the classification of unobservability is theory dependent, it is wise to get beyond the slogan that it is theory dependent and see, case by case how it is theory dependent.

The differences in these two cases of unobservability indicate that general claims about unobservables hide important information. Regarding the existence or nonexistence of the unobservables, for example, a Robertson-Walker cosmology requires that the objects which it classifies as unobservable exist. But elementary particle physics is not inconsistent without the existence of quarks, as it would be without the existence of neutrinos to account for energy and momentum conservation. It is just that quarks allow simple, balanced explanations. The differences in the nature of their unobservability tend to impeach any general claims about the actual or proper attitudes one does have or should have as to the observability or existence of the entities of quantum mechanics. And if it is unwise to generalize across quantum entities or across the class of things unobservable in principle, it is the more unwise to generalize across all unobservables.

2. THINGS UNPERCEIVABLE IN FACT

Consider now four cases of observability claims in which the object in question can interact and convey information through a machine but not to a human body directly. The machine must be able to report the information of interaction in a way that is meaningful to the scientist reading the output.

In pursuing these examples of things unperceivable in fact, attention should be focused on the conveyance of information from the object to the observing scientist. It is the relevant physical theories which are being asked to rule on the possibility and nature of interaction and to process the information, so it is appropriate to look carefully at the ruling to explicate the complexity of the informational processing. It is important too to assess the amount of independence between the theory of observation and the theory of the object.

The quantum mechanical state function, ψ

The observability status of the state function presents something of an enigma. It is not something that can be read in a straightforward way in text books on the subject, because there are comments on the one hand that "there is no experiment which will measure ψ at a given point at a given time" (Heisenberg 1949, 51), and references on the other hand to states which "can be directly observed," (Schlegel 1980, 210), and claims that "a state is a measurable quantity" (Jauch 1968, 93). I plan to sort out these comments by pursuing two questions. Is the state function the kind of thing which it even makes sense to classify as observable or unobservable? And is there any observation process one can perform on a quantum system which produces sufficient information to infer the state of the system?

To understand the principles of quantum mechanics, and in particular to faithfully interpret the state function, it is essential to distinguish between three kinds of quantum mechanical system, the individual, the pure-state ensemble, and the mixture. An example of an individual system would be a single particle, say, in a double-slit experiment. A large collection of particles all in the same circumstances, in the same double-slit apparatus or the same Stern-Gerlach device, would make up an ensemble. The ensemble is homogeneous in the sense that no subensemble can be found within it which evolves differently from the rest. A mixture of the systems is an inhomogeneous collection in which subcollections may evolve differently.

The state function ψ is a property of an ensemble, not an individual or a mixture. It is a complex valued function which, by hypothesis, contains all the information such as position, energy, spin, and so, of the ensemble. When an ensemble is in the state ψ, it is sometimes said that each individual system of the ensemble is itself in the state ψ. This is how to make sense, for example, of the double-slit experiment in which particles are sent through the slits one at a time. But this is a heuristic extension of the concept of the state. Strictly, the state function is defined only for an ensemble. For the object-property approach to observability then, the appropriate formulation is to ask of an ensemble (the object) whether it can interact to convey the information that it has the property of being in the state ψ. What we are looking for is information which comes from the ensemble and reports its quantum state.

In general ψ is a function of time, $\psi(t)$, which carries the system-information continuously through time. The evolution of $\psi(t)$ is described with a wave equation such as the Schrodinger equation. But unlike other wave equations in physics like a water-wave equation or an electromagnetic wave equation, the value of ψ has no obvious physical interpretation. With water waves, the value of the function which is described by the wave equation is simply the vertical displacement of water as measured from the undisturbed position. That is, the value of the water wave function is a position, and to observe the wave one observes positions of the water surface. By observing several values of vertical position at different horizontal locations or at different times, one can infer the wave function. In this way it can be said that water wave is observable by virtue of its interacting in a correlated way with a human apparatus.

With an electromagnetic wave, the value of the wave function is the electric field vector **E** and the magnetic field vector **B** combined. These fields are responsible for producing forces on a test charge which can be part of an observing apparatus. So the values of the electromagnetic wave do interact in a correlatable way and so may be unobservable in fact but not in principle.

But the value of the quantum mechanical state is not obviously associated with any physical property. Such quantities as position and electric charge are drawn out of ψ by using the appropriate operator, but none of these is the value of ψ. So one might wonder why the observability of ψ should even be considered.

There are two forms of information in the resulting apparatus state after interacting with the quantum system, individual eigenvalues like position or spin orientation, and probability distributions of these values. Insofar as the state function describes an ensemble it is the measure of probabilities which are fundamental and more likely to contain information of the state. Probabilities always involve squared terms of the state function, $\psi\psi^*$. If the ensemble is in the state ψ, then the probability that measurement of some observable A will result in a value a is $|\langle u|\psi\rangle|^2$, where u is the eigenvector associated with the value a. In the coordinate representation, $\psi(r)$, for example, the square of the state function can be regarded as a position probability. That is, $|\psi(r)|^2 \cdot d^3r$ is the probability that if the position of the system is measured, it will be within the region d^3r. By considering the square value $|\psi|^2$, only real values are used and it is possible to associate $|\psi|^2$ with real, measurable quantities in the world. So one might want to ignore ψ and consider only $|\psi|^2$ as a fundamental quantity. But this cannot be allowed because the superposition principle which is one of the most important features which distinguishes quantum from classical physics, requires that systems interact by combining states ψ and not probabilities $|\psi|^2$. So even though it is only $|\psi|^2$ which has a physical interpretation in the sense of being manifested through interaction with a classical apparatus, ψ itself has an essential role in the description of interaction of quantum systems.

Probabilities are observable as relative frequencies. (Whether probabilities simply *are* relative frequencies is another question, but surely relative frequencies are a manifestation of probabilities). The trick is to use an observing apparatus to correlate the quantum mechanical characteristic in question to some parameter of the apparatus which is simply perceivable. A good example is the Stern-Gerlach apparatus in which an inhomogeneous magnetic field correlates the spin orientation of the particles to the position on the screen where they subsequently land. Probability of spin orientation in the ensemble shows up as relative frequency of hitting the screen at a certain point. In this way, the apparatus-state contains information on squared terms of the state, $\psi\psi^*$. The question is, can one infer the value of ψ from the squared terms, the probabilities?

To answer this, add one more detail to the quantum mechanical description of the state function. In general, the state function can be expressed as a superposition of eigenstates of the representation one chooses. For example, if one chooses the coordinate representation then ψ will be expressed as a superposition of eigenstates of the position operator. In general, $\psi = \sum c_i \phi_i$ where the c_i are complex coefficients and ϕ_i are the eigenstates of the chosen representation. $|\psi|^2$ can be normalized such that $\sum |c_i|^2 = 1$. To know the value of ψ means to know the values of the c_i.

It is easy to get values for $|c_i|^2$. These come from the probability measurements, the squared terms $\psi\psi^*$. If a_i is the eigenvalue associated with ϕ_i, measure $\mathrm{Prob}(a_i \mid \psi) = |\langle\phi_i|\psi\rangle|^2 = |c_i|^2$. This gives the real part of c_i but not the imaginary part which accounts for the relative phase between the components of ψ, and hence for interference effects. The unnesting of information about relative phase is best described using an easy, but realistic example.

Consider a two-level case, in particular an ensemble of spin-1/2 particles where ψ can be regarded as a superposition of states of z-component of spin, $\psi = c_+\psi_+ + c_-\psi_-$. Since the coefficients are complex, write them write them as $c_+ = r_+ e^{i\phi_+}$ and $c_- = r_- e^{i\phi_-}$. Multiplying by a global phase factor causes no change in the state, so transform $\psi \rightarrow e^{-i\phi_-}\psi$, and ask about the new state function $\psi = r_+e^{i\delta}{}_+ \; r_-\psi_-$. δ is the relative phase difference $\phi_+ - \phi_-$ which is significant for interference effects. The question of observability boils down to finding values for r_+, r_-, and δ in the final apparatus state.

The ensemble is a large, homogeneous collection of systems. We can measure on *part* of the ensemble the probability of z-component of spin being up, $\mathrm{Prob}(S_{z+} \mid \psi)$ using a z-oriented Stern-Gerlach apparatus. This gives

$$r_+ = [\mathrm{Prob}(S_{z+} \mid \psi)]^{1/2}.$$

Then r_- follows from normalization,

$$r_- = [1 - r_+^2]^{1/2}.$$

On *another part* of the ensemble, measure the probability of the x-component of spin being down, $Prob(S_{x-} \mid \psi)$. Then from the calculated expansion of the (S_{x-}) state in the S_z representation,

$$\psi_{x-} = \frac{1}{\sqrt{2}}(\psi_{z+} - \psi_{z-}),$$

it follows that

$$Prob(S_{x-} \mid \psi) = [r_+^2 + r_-^2 - 2r_+r_-\cos\delta\ ].$$

Solving for δ,

$$\delta = \cos^{-1}\left\{\frac{1 - 2\ Prob(S_{x-} \mid \psi)}{2\ r_+r_-}\right\}$$

The physical setup for this case of observing ψ is pictured in figure 7.

The parts of the ensemble which have been measured, either for $Prob(S_{z+} \mid \psi)$ or $Prob(S_{x-} \mid \psi)$ are no longer in the state ψ. Measurement turns pure states into mixtures. (Accounting for this is the task involved in dealing with the so-called measurement problem. It is not directly an issue for the discussion here). Sampling from the ensemble as we have, measuring $Prob(S_{z+} \mid \psi)$ on part, and $Prob(S_{x-} \mid \psi)$ on part of what remained, gives enough information to infer the state of the unsampled part of the ensemble.

This kind of procedure for determining ψ can be extended to higher dimensions. In general 2N-2 measurements are necessary to determine the coefficients of in a N-dimensional representation (2N because each of the N coefficients has 2 degrees of freedom, magnitude r and phase ϕ, but -2 because one degree of freedom is constrained by normalization and a second by factoring out a global phase term). Cases of infinite dimensionality such as $\psi(x)$ of a free particle can be done with measurements of the distributions Prob(x | ψ) and $\frac{\partial}{\partial t}$\{Prob(x | ψ)\}. This procedure is described by Kemble (1937, 71), and by Blokhintsev (1968, 45).

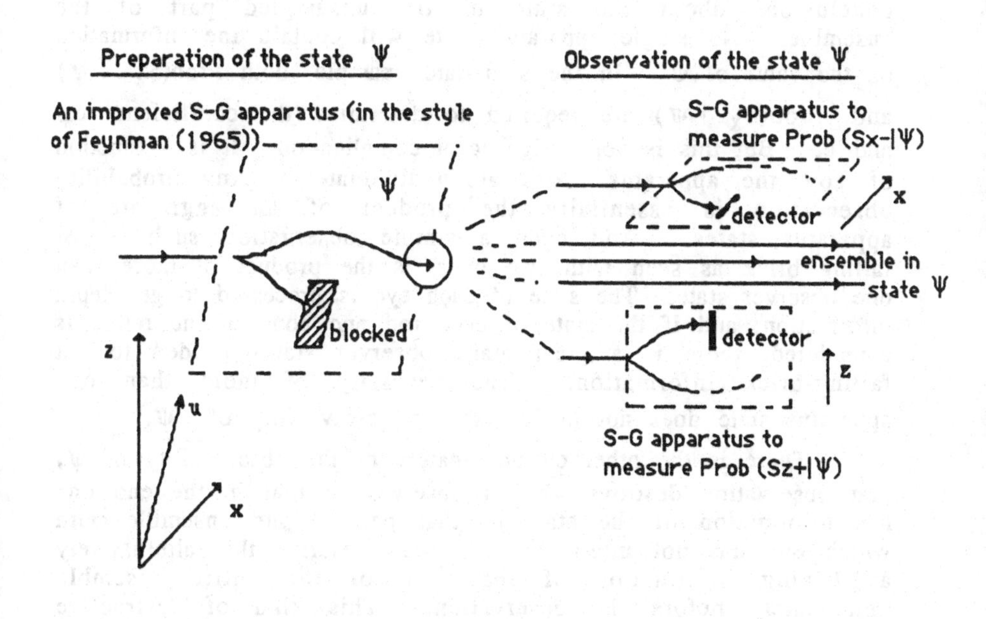

figure 7

The important aspects are the same in all cases. One starts with an ensemble, a pure state ψ. In practice this must be prepared. In the case of the spin-state, an ensemble can be prepared with an inhomogeneous magnetic field at some random angle, as in figure 7. There is a temptation to say that preparation of the ensemble amounts to observation, since knowing the orientation of the magnetic field is information of the state. But it is not observation in the interaction-information sense because the causal process is reversed. The chain of interaction is not from the ensemble, through the magnetic field, to the observer as it would have to be to be information *from* the object. Rather it is a causal influence from apparatus to object. The information originates in the apparatus. One prepares the state and can thereby infer the value of the state, but it then still remains to observe the state that has been prepared, observing it by being affected by the ensemble.

The observation is done by sampling, and in the process destroying, several parts of the ensemble and drawing conclusions about the state of the unsampled part of the ensemble. No single apparatus state will contain the information on the value of ψ. In the spin-state example, both Prob(S_{z+} | ψ) and Prob(S_{x-} | ψ) are required to determine the coefficients c_+ and c_-. But this is not a significant complication. It is a function of how the apparatus states are individuated. Any probability observation is essentially the product of an aggregate of apparatus states. And even a simple observation such as of falling brick as seen with two eyes, is the product of more than one observer state. The state of each eye is processed to get depth information, and if the state of each rod and cone in the retina is considered, there is a great many observer states needed to get falling-brick information. The necessity of more than one apparatus state does not distinguish the observability of ψ.

There is the other curious feature of the observability of ψ, that observation destroys what it observes so that in the end one has information of the state of that part of the ensemble with which one has not interacted. On can describe this alternatively as having information of the state of the entire ensemble immediately before the observation. This kind of destructive observation is not unique to the case of observing ψ. To measure the chemical composition of a rock sample, for example, one might have to pulverize and dissolve some part of it to observe chemical characteristics, and then generalize the properties of the now destroyed part as being characteristic of the whole. We observe the light of the sun, to take another example, by absorbing a small fraction of its production and claiming information about the light as it was before absorption. Destruction is commonplace in observation.

The conclusion then is that a quantum ensemble can interact with an observing apparatus in such a way that the information that the ensemble is in the state ψ is contained in the state of the apparatus and is intelligible to a human observer. The quantum mechanical state function of an ensemble is not unobservable in principle but is at worst unperceivable in fact. The state function of an individual system, on the other hand, is unobservable in principle. This is because determination of ψ requires more than one measurement of probability, and the first such measurement alters the state, precluding any subsequent measurement on the initial state. In the example of spin-state, if there is only one system, a measurement of $\text{Prob}(S_{z+} \mid \psi)$ is information of r_+ and r_-, but it is also a destruction of the state ψ, hence there is no way to get information of δ. (The problem is actually worse than this since a measurement on an individual system results in an eigenvalue, that is $S_z = +1/2$ or $S_z = -1/2$, but not a probability. There is no way to get even the information of r_+ or r_- from the eigenvalue.) But since states are, strictly speaking, attributes of ensembles and not of individuals, it is no detriment to the observability of states that the individual state turns out to be unobservable in principle.

The fact that observation of a quantum state requires multiple sampling, both in the sense of measuring probabilities and of having to measure a variety of probabilities, means that the state cannot be observed at a single spacetime location. Hence we can reconcile Heisenberg's comment, quoted at the beginning of this section, "there is no experiment which will measure ψ *at a given point at a given time*" (1949, 51, emphasis added). This is true, but it does not mean that ψ cannot be observed in the way described above.

To summarize the observability of ψ it can be located along the dimensions of observability. The state of a quantum ensemble is unperceivable in fact but not unobservable in principle. The Stern-Gerlach example demonstrates how an apparatus can correlate ensemble state information to apparatus state information. To evaluate the directness of such an observation, consider for each probability measurement, the interaction between ensemble and the magnetic field of the Stern-Gerlach apparatus and between ensemble and screen used to record relative frequencies. Clearly, directness is different in different cases, depending on the dimensionality of the state-space. The 2N-2 required measurements indicates the number of required interactions and hence the measure of directness of the information acquisition. The relevant quantum theory describes the number and nature of interactions required.

The amount of interpretation does not necessarily depend on the dimensionality of the state-space. The different measurements required could be, as in the case described, simply rotated tokens of the same type. In such a case, all the measurements are described and accounted for by the same aspects of quantum theory. On the other hand, we could imagine cases in which the 2N-2 measurements were significantly dissimilar and required us to draw on a variety of aspects of quantum theory. The point is that evaluation of this amount of interpretation required to acquire information of a state function will be a case-by-case operation.

As for independence of the interpretation, there isn't much in any case. It's all quantum mechanics. The interpretation of relative frequency observations and the calculations of the coefficients c_i in the 2-level case described are both axioms of quantum mechanics and are virtually part of the definition of the state function. So, though the relevant physical theory describes the <ensemble, state function> pair as being able to interact and convey information, it is important to realize that the relevant theory is quantum mechanics itself, reporting on its own observability.

The photon

The question of the observability of the photon is as enigmatic as it is important. Interaction between photons and humans is certainly an important part of many observations, but that does not imply that the photons themselves are observable. A group of photons may be able to convey the information that, say, this page is rectangular, yet be unable to impart any information about any one of the photons themselves. And it is the acquisition of information about x which is necessary to claiming that x is observed. This raises the general question of whether a means of observation must itself be observable, or at least as observable as the object being observed. It is a question relevant to Shapere's (1982) discussion of observing the interior of the sun. His focus is on using neutrinos to observe the sun and not on the observability of neutrinos themselves. Pursuing the case of photons can be expected to suggest an answer to the general question.

Unquestionably photons interact with the human body. As I sit here, light from the page enters my eye and is absorbed by the pigment in the retina. Photons of infrared frequency from the radiator arrive at the skin and exchange energy, while cosmic gamma rays are streaming in through the roof and altering a chromosome here and there in the cells. The tan color of the skin is evidence of a summer spent interacting with photons from the sun. But can any of these events be classified as observations of photons? Given that interaction occurs, the important factor is the meaningfulness of the interaction. Do I get information about a photon in any of these processes? A photon is neither so clearly observable as a moon of Jupiter nor so clearly unobservable as an electron. It is rather one of the vague cases which van Fraassen claims we can live with.

Foss (1984, 84) recognizes the importance of photon observability. Photons are, after all, elementary particles and any general claims about the unobservability of microparticles must include photons. If one admits, based on the physiological fact of the human eye being a photon detector, that photons are observable, then one is obliged to explain why other microparticles are not observable. The body can detect electrons, for example, as in a static shock from walking over a carpet. And if photons are not observable, then there must be something more to observation than detection, that is, there must be some meaningful information such as "that it is a photon" or perhaps "that it has energy (spin, charge) ----". Foss opts to generalize over microparticles and declare them all observable, as observable in fact as mice (1984, 90). He concludes this from a description of observing an object which equivocates between being "informed *by* it" and getting "some information *about* it" (page 90, emphasis added). He places greater importance on the causal influence of the object on the observer than he does on the nature of the information. But there are important epistemological differences between information by an object and information about the object. I get information by radio, by the arrangement of transistors, diodes, capacitors, etc., each morning, but in the process I do not necessarily get information about the transistors etc.. A careful analysis of observability in such vague and important cases as the photon must be sensitive to the content of the information imparted in the interaction. To such an analysis we now turn.

First it is necessary to be clear on what a photon is in order to know how one can be observed. As a quantum of electromagnetic energy, the photon is distinguished from being merely a short but continuous pulse of energy. This means that a single photon interacts with a single electron. It cannot spread its influence over several electrons as could as unquantized pulse. This is a crucial distinction between the classical and quantum mechanical descriptions of radiation, and as such it is an essential property of the photon. For this reason, specify the evaluation of observability to individuated photons, and look for information not of groups but of individuals. This is an important restriction because the immediacy of observability of ensembles of photons is likely to be different from the immediacy of observability of individual photons. The focus here is on the observability of individual photons because therein lies the immediate relevance to quantum mechanics.

To be faithful to the plan of assigning observability to an ordered pair of the form <object x, property P>, it is helpful to point out specific properties of photons which might figure in observation. An observation interaction must happen at a location and can potentially report the position of the photon. That is, a photon has the property of a spacetime position which can be reported in terms of being present at the observer and at the time of observation. This is likely to be the easiest observation, to claim that a photon is present. Other properties which could be the basis of observing a photon include its energy (frequency, wavelength), momentum (including the direction of motion), spin, electric charge and mass. To observe a photon, interaction must occur such that one is informed of some one of these properties of the photon.

Can such informative interaction take place between photon and the body? To answer this it is necessary to appreciate just what it is the eye can do. Only then can one know whether it is true, or rather, what is meant, when a particle physicist proclaims, "One of the massless particles, the photon, we see everyday" (Ford 1963, 113).

Consider the interaction between a photon and the eye and see if the information "that a photon is present" is conveyed. To interact with the eye the photon must be absorbed, and this event takes place in the pigment of a rod or cone in the retina. Absorption of the photon initiates a series of chemical changes in the pigment resulting in a change in the permeability of membranes in the rod or cone to ions. Alteration of the permeability to ions has an effect on the charge separation within the rod or cone and in this way leads to an electric current being passed along nerve cells to the brain.

What information of itself can an individual photon impart through interaction with the eye? In discussing the threshold of optical stimulation, only the rods are worth considering, for they are the more sensitive of the two kinds of receptors. Using a statistical correlation of experimental data between the estimated number of photons reaching the rods and the detection of light, Brindley concludes that "one quantum must suffice to stimulate a rod" (1970, 187). But stimulating a single rod is not necessarily detecting, let alone observing, and Brindley goes on to say that it is unclear how many photons are required for a signal to be sent to the brain. Again using a statistical analysis of experimental data, he reports that a mean number of five photons being absorbed is necessary to detect a flash of light. That is, a mean number of five photons is needed to impart the information "that there is a flash of light". This, of course, does not rule out the possibility that a single photon could initiate the signal and be detected as a flash of light, though the probability is of the order 10^{-4} (Brindley 1970, 190).

In terms of the concept of informational content as described in chapter two, the observer state of perceiving a flash of light cannot be said to contain information of an individual photon (nor of individual photons). The signal from the eye does not unambiguously report the arrival of one photon, or of any specific number of individual photons. It does not provide information on individuated photons, quanta of light, any more than standing in the rain offers information on individual water molecules. Even with background knowledge k which includes the quantum theory of light, the experience of a flash of light does not license the claim, "that's an individual photon". It could be any number of photons, and whereas the actual number of photons present is not so important, being able to individuate quanta of light is. Photons are particles of light, and unless the experience indicates a particle-like quality, it does not report on individual photons.

Apparently then, insofar as the observing apparatus is restricted to the human eye, the accurate assessment of photon observability is that whereas a photon can interact with the observer, it cannot do so in a meaningful way. The interaction does not convey any information of the sort "that the photon is ____," that is, no information of the photon. Assuming that the eye is the body's most sensitive photon receiver and hence its most likely photon observer, the inability of the eye to observe the photon can be generalized to an inability of the human body.

Given that the individual photon is unperceivable to a human, it remains to be shown how photon information can be communicated to machines and thence to the scientist. One can learn about the process of observation by tracing the conveyance of information in a sequences of interactions which a scientist might refer to as "observing a photon". Different sorts of machines are used to interact with different energies of photons. Electric circuits as in radios are used to encounter low energy photons, while temperature sensitive devices respond to more energetic, infrared photons, and photoelectric devices probe for the still higher energies. Since it is in the high energy photons that the quantum effects are most easily identified, that is, where photons are most readily individuated, it is there that observing photons is most interesting, and it is there that I will focus. In any case though, high energy or low, the interaction between photon and apparatus must, on the individual scale, be between photon and some charged particle in the apparatus. This initial event is described by QED. It is what follows, the rendering of information from this event, that is different for different domains of energy.

High energy photons interact and pass information by photoemission of electrons in the detector. In the process the incident photon is annihilated and all of its energy is transferred to the electron. Some of the energy is used to free the electron kinetic energy. So there is a correlation between the photon energy $h\nu$ and the electron energy E:

$$h\nu = E + \text{binding energy}.$$

The one electron which is liberated by the one photon is accelerated in an electric field in a photo-multiplier apparatus and made to collide with other bound electrons, releasing some of these electrons, the number released being correlated to the energy of the initial electron. These secondary electrons are themselves accelerated such that they collide and liberate more electrons, the process being repeated a dozen or more times such that a cascade of free electrons is liberated and can be collected at an anode, resulting in a pulse of current which is measurably large. It is the current pulse then, displayed on an oscilloscope or as the change of a digital pulse counter, which reports the presence of a photon. And since the strength of the pulse is proportional to the energy of the photon, one is presented with information not only of the form "that a photon is present" but also "that the photon has energy $h\nu$". This information is passed from photon to primary electron in the initial interaction, then from primary electron to the group of secondary electrons, and so on to the circuitry and the counter. The final electrons which arrive at the anode never interacted with the photon. They are a dozen or more events distant from the photon, yet the information of the photon is present in the group. No single one of the final electrons carries photon information since the relevant datum is a function of the number of electrons collected.

There are several criteria with which to evaluate this passage of information about the presence and energy of a photon. For one thing, this is a case what Shapere would call indirect observation of the photon because the messenger, the participant in the final interaction which brings the information to the scientist has not itself interacted with the object. As pointed out in chapter two, the identification of *the* messenger is quite arbitrary and it is more informative to assign degrees of directness which reflect the number of times between object and observer that the information must change hands. In this spirit, photon observation is relatively indirect, relying as it does on a bucket-brigade of information being passed from group of electrons to group of electrons. The directness in this case is indicated, in part, by the number of stages in the photomultiples. It is interesting to note though that all of the stages in the cascade of electrons are the same in the sense of being accounted for by the same laws.

What is the amount of physical theory needed to account for the information? The exercise here is to see how physical theory decides on observability and this entails revealing what theories are invoked in the process and how they are used. In the case of the photomultipler, the theories supporting the observation are few and straightforward. An elementary understanding of the photoelectric effect shows that the light quantum interacts with a single, bound electron and frees it with an amount of energy dependent upon the energy of the incident photon, accounting for the binding energy. Electrostatics and the basics of electron collisions attest that the primary electron imparts it's energy and the information it bears to the secondary electrons, and they to the tertiary, and so on to the final electrons and the anode. The rest is electronics. The great indirectness of the process of cascading electrons does not yield a great amount of interpretation since the many interactions are described by a single set of laws.

The other important feature of the account of observing a photon is the independence of interpretation. In observing the interior of the sun, the case detailed by Shapere (1982), one had to do much chemistry and neutrino physics to find the information about the sun. In the solar case the theoretical account of observing, the chemistry, etc., is quite distinct from the theory of the sun. Independent theories are invoked as auxiliaries for observing the sun. But in the case of the photon, the photoelectric effect is a key link in the account of observing, yet photons are introduced in physical theory in part to explain the photoelectric effect. So this is observing an object by observing an effect for which the object is the best explanation, and it is an important proviso to add to photon observation reports that a vital part of the theory used to license the claim that the final information is information of a photon is a basic part of the theory of the nature of the photon itself.

Summarizing the observability of photons, it is deceptive to claim, on the basis of human eyes being photon receivers, that photons are observable. Such a classification ignores the important distinction between interacting with an object and interacting in a way that is informative of the object (not simply informative *by* the object). Just as standing in the rain cannot be described as observing water molecules, gazing at the sky is not observing photons. Photons, and their properties studied here, position and energy, are therefore classified as unperceivable but only contingently so. There are instruments which, as described by relevant theory, interact with and inform of photons. The relevant theory, it must be kept in mind, is not at all independent from the theory which describes the object.

Examples of observing in which this close dependence between theory of the object and theory of observing does not exist are easy to find. The next example to be discussed presents such a case.

Observing with an electron microscope

In studying the variety of examples of observability claims made by the working scientist, it is always motivating to run across statements like the following. "[The electron microscope] brings man's ability to 'directly' observe and study matter down to the molecular realm" (Wischnitzer 1970, 1). For the interaction-information account, this is an invitation to detail the mechanisms behind the observation with an electron microscope and to understand why this can be called 'direct' observation, where the directness must be mitigated by quotation marks. Is electron-microscope observation, which can resolve objects of 0.1 mn size, any less direct (or less 'direct') than optical-microscope observation which resolves to 100 nm? To answer this, one must sketch the essential features and trace the communication of information through an electron microscope much as Hacking (1981) has done for the optical microscope.

Just like its optical counterpart, the electron microscope is designed to exploit several different features of the electron to convey information of the specimen. Which property of the probing electrons is to be used, and the details of incidence and detection are decided according to the type of specimen, whether it is a thin, nearly electron-transparent organic tissue, for example, or a thick, dense crystal. The earliest and most straightforward device for imaging a specimen according to its electron interaction features is the conventional transmission electron microscope, or CTEM to those in the know. There are two ways to produce an image of the specimen using a CTEM, one using an incoherent source of electrons, the other using a coherent source.

With an incoherent source, a CTEM works in the following way. A source of electrons is created by heating a piece of metal, that is, by thermionic emission. These electrons are accelerated in an electrostatic field and are focused into a narrow beam by passing through a magnetic solenoid, usually referred to as the condensing lens. Typical electron energies are 100 keV to 1 MeV. The beam of electrons hits the specimen where some of the electrons are elastically scattered by atomic nuclei. Many of the scattered electrons are deflected by large angles from the uninterrupted beam direction, such that a small aperture placed just down-beam of the specimen will block these scattered electrons and allow only undeflected or slightly deflected electrons to pass. The electrons which pass through the aperture are then magnetically focused such that electrons from a particular point of the specimen are focused to a single point at the image plane, as is shown in figure 8a.

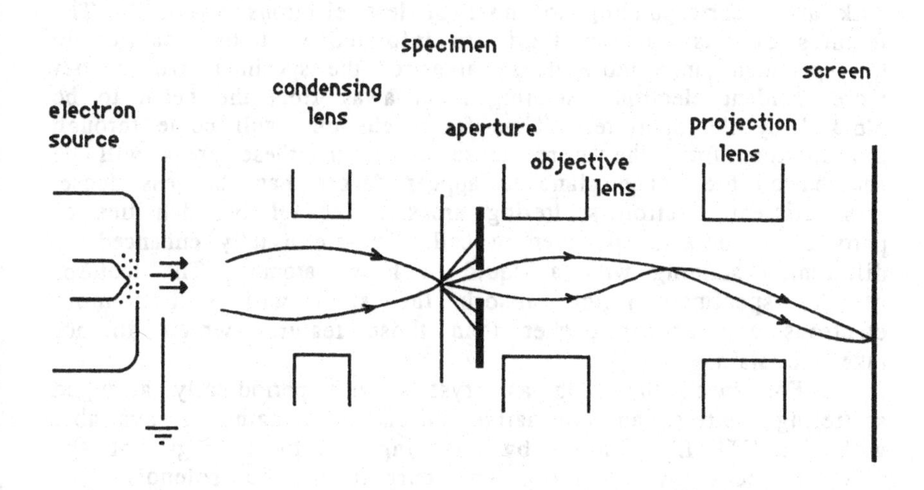

figure 8 (a)

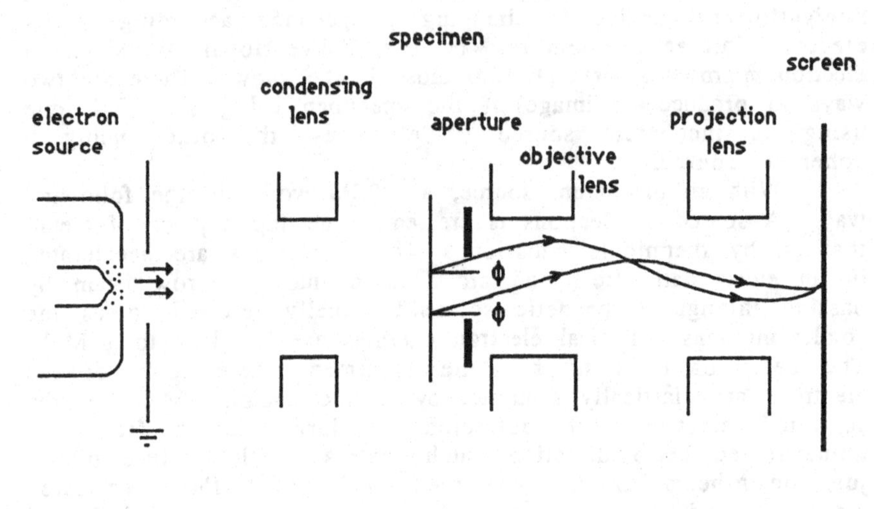

figure 8 (b)

At the image plane the electrons strike a phosphorescent screen or a photo emulsion plate creating a contrast field of light and dark areas corresponding to more or less electrons arriving. The features of this contrast field are informative of the features in the specimen since more dense areas of the specimen will scatter more incident electrons, sending them away from the beam to be blocked by the aperture. Thus fewer electrons will come through the aperture from the more dense areas and these areas will be focused on the image plane to appear darker than the less dense, less efficient electron scattering areas. The relative densities of parts of an organic specimen, a cell, say, are usually enhanced by differential staining with a liquid of heavy atoms. The features of the specimen which absorb the stain will scatter more electrons and appear darker than those features which do not take the stain.

For specimens such as crystals with periodically arranged scattering centers, an alternative means of imaging is available with the CTEM. Simply by changing the focal length of the objective lens by changing the current in the solenoid, the electrons which leave the specimen at the same angle ϕ regardless of where on the target they are scattered, can be focused to a point at the image plane (see figure 8b). With this arrangement the CTEM exploits the wavelike properties of the electrons and images the Fraunhofer diffraction pattern of the specimen. The form of the image is an array of dots whose relative positioning corresponds to the geometry of scattering centers in the crystalline specimen.

The wave qualities of the electrons can be further utilized if the incident beam is coherent. This requires that the electron source produce electron waves which are at the same phase, and this is usually done by field emission, literally pulling electrons from a fine metallic point with a strong electrostatic field. These phase-coherent electrons are targeted on the specimen as before and these that are scattered are phase-shifted by an amount δ which depends on the scattering potential of the target. (Particle scattering always produces a change of phase.) The scattered electrons also have a longer distance Δd to travel to the image plane than do the unscattered electrons. The objective lens is tuned to reunite scattered and unscattered beams from each point in the specimen, producing an image point with an intensity which depends on the amount of constructive or destructive interferences between the phase shifted $\delta + 2\pi \frac{\Delta d}{\lambda}$ electrons and the unshifted beam. This is pictured in figure 9.

Accounting for the complicated interplay of lens aberration, lens focusing point, diffraction and interference between the many scattering centers, the contrast field produced at the image plane reproduces the scattering potential field of the specimen. This use of the electron microscope is known as phase contrast imaging.

The CTEM uses an electron beam which is wide enough to cover the entire area of interest in the specimen at once, just as an optical microscope illuminates the entire specimen being viewed. A scanning electron microscope, on the other hand, operates with a finely focused electron beam (cross sectional area on the order of 0.1 to 1.0 nm) which is incident on only small spots on the specimen at a time. The beam sweeps back and forth across the specimen and at each point the relevant electron properties (absorption intensity, phase contrast, etc.) are detected. Correlating information about the position of the beam and the resulting scattering, an image is formed on a television screen. For thick, dense specimens where most of the beam is scattered at the surface, information is detected in the form of the number, energy and direction of secondary electrons released at the specimen surface. These data report on surface geometry. Characteristic x-rays emitted when specimen atoms relax after having inner electrons scattered out report the elemental make-up of the specimen.

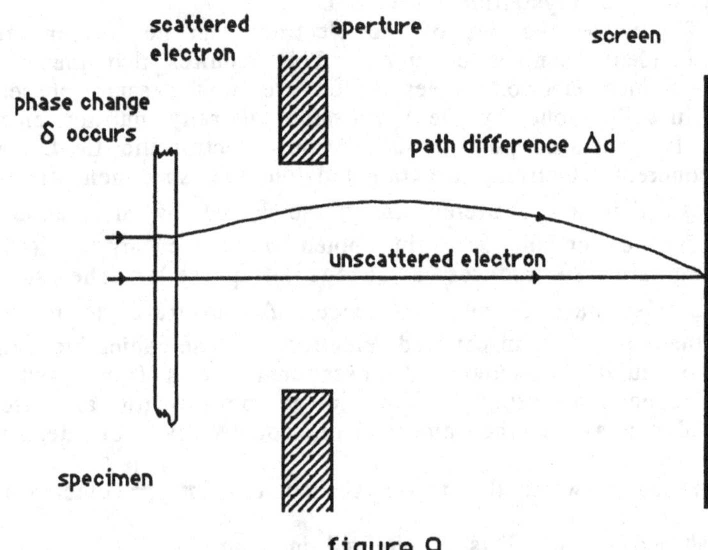

figure 9

With thinner and less dense objects one can analyze the electrons which pass through the specimen using a scanning transmission electron microscope, an STEM. This functions much as the CTEM except that the image does not appear by phosphorescence at the image plane but by signals from an electron detector and the beam position, correlated and displayed on a television screen. As it is most typically used, the STEM differs from the CTEM in that it uses dark field imaging. Instead of detecting the electrons which pass through the circular aperture, a dark field image is produced by detecting the electrons which do *not* pass through the aperture. It counts those electrons which are widely scattered. This is done with an annular-shaped detector which reports many electrons when the incident beam is hitting a dense scattering center, and a few electrons when the beam is not so heavily scattered. By ignoring the undeviated central beam the dark field imaging does not get information of phase difference between scattered and unscattered electrons.

Having outlined the basics of electron microscopy, we can consider two specific examples of things seen through electron microscopes. Consider first a CTEM, phase contrast image of a DNA molecule. Specifically, take the case of being informed by a CTEM image that looks like figure 10, that the molecule is a closed loop.

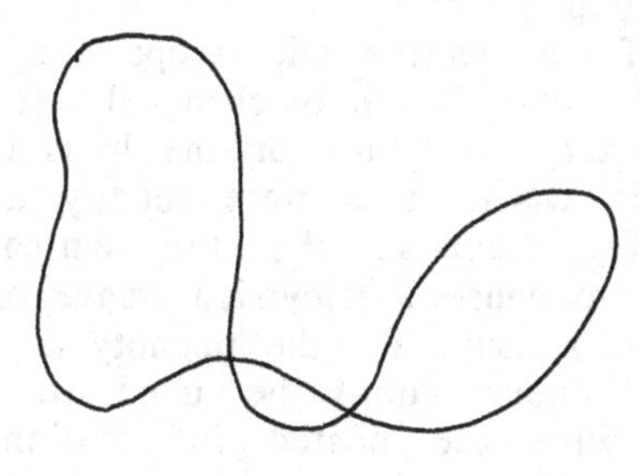

figure 10

What one sees in this case is a simple pattern of dark lines on a lighter background. The reliability of the information, the assurance that this image corresponds to the shape of the specimen, is founded on an understanding of electromagnetism (to account for the emission of electrons and the effects of the lenses), the quantum mechanics of electron scattering, the chemistry of the stain used, the optics of diffraction and the fluorencense of the screen. These are the physical theories which sanction the claim, "that is a strand of DNA, and it is a closed loop."

Much of this theoretical support is likely to be forgotten if what you are interested in is DNA. This is because the information is displayed by the CTEM in an easily accessible way such that even someone unfamiliar with the mechanics of electron microscopy can quickly notice that the molecule is a closed loop. Contrast this with some biochemical test which, by some phenomenon of micro-organisms, indicates the same thing, namely that their DNA molecules are closed loops. Lehninger (1970, 650) describes such a procedure which indicates that the DNA molecule in E. coli is structured as an endless, closed loop. The information is obtained in experiments by interrupting the poor E. coli at various stages of conjugation and noting the type and number of phenotypic traits in the progeny. In this case a glance at the results is not enough to lead one to the belief that the DNA is a closed loop unless one has a good idea of the whole scientific account which links a closed loop of DNA with the phenotypic manifestations. There is more theoretical preparation necessary to noticing (to borrow Hacking's term) that the strand is a closed loop in the biochemical test than with the electron microscope screen.

It seems as if the electron microscope image is more a case of observing the DNA than is the biochemical test, but it would be misleading to draw this distinction on the basis of the microscope screen presenting the DNA in a more readily noticed form. It would be misleading because the biochemical results could conceivably be made to cause a television image of the closed loop which is theorized to account for the phenotypic results. That is, some form of transducer could be used to relay phenotype information, along with the nested information of the closed loop, into a machine which produces a picture of a closed DNA loop. The information of closed-loop DNA still arrives at the screen and is now easily noticed. Noticeability cannot sanction distinctions of observability in these cases. Rather, it is the manner in which supporting theories are invoked in producing the final image which divides the cases. Where biochemical analysis is used the physical laws which describe the passing of information from strand to image are, naturally, biochemical and genetic laws, laws in which DNA and the effects of closed-loop DNA must figure. But with the electron microscope the laws which trace the information are indifferent to DNA, its structure or the effects of its being a closed loop. Here is an opportunity to use the precise description of independence developed in chapter two. In the biochemical case, T_x2, that part of the theory of DNA used in the account, will include a claim of the form "if the DNA is a closed loop then ___________ (phenotype property)." This will be a case of T_x1 and T_x2 not disjoint and T_x2 is a member of $\{T_i\}$. In the microscope case, on the other hand, T_x2 must include claims about the composition of the molecule, to account for its differential absorption of stain and its scattering of electrons, but no mention of the shape of the molecule. Hence, in this case, T_x1 and T_x2 are disjoint and there is a significantly higher degree of independence than in the biochemical report.

The conclusion we can draw is that it is not a matter of more or less theoretical support necessary to account for the information, nor the image being more or less noticeable to the untrained, which are relevant to the degree of observability in this case. Rather, it is a question of more or less independence between the laws of observing and the theory of the object.

A second specific example of electron-microscope observation that will further motivate this conclusion is based on an account of "Electron Microscopy of Individual Atoms" (Isaacson, et al 1979, Chapter 13 title). The microscope described by Isaacson is an STEM with a beam so narrowly focused that it scatters off one atom at a time. This eliminates effects of interference from multiple scattering centers. The imaging used is dark field imaging, hence there are no phase coherence effects. The informational flow is quite simple in that an annular detector collects many scattered electrons when the beam is incident on a massive scattering center. The signal is sent to a television screen and in a position corresponding to the position of the incident beam, a white dot is produced which is the image of the atom responsible for the scattering. In this way, crystal arrays, or even single atoms when they are alone in a crowd of much lighter atoms, can be shown. There are photographs of these kinds of results in Isaacson (1979, 359). The picture of a "single mercury atom" (Figure 13.8a in Isaacson) is completely described as a white spot on a cloudy dark background. The authors present these results in a section with the cleverly ambiguous title, "How to Visualize an Atom".

One can ask why this result is not described as the image of a nucleus rather than an atom. It is the nucleus, after all, which acts as the massive scattering center. The point is that the white spot more immediately represents the presence of an electron scattering center than it does an atom, and calling it information about an atom relies on a theory of atomic structure. The STEM probe and imaging are essentially just a compact system of Mott scattering (which is just Rutherford scattering with electrons rather than alpha particles as probes) where scattering information is correlated to the position in the specimen. This seems no more direct an observation of the atom than does some other phenomenon, like beta-decay, which is explained by atomic theory in terms of the presence of an atom. Both a beta-decay event and the location of a massive electron scattering center present the information "that an atom is present" but in both cases the information is rendered only with the help of atomic theory. Imaging a single atom by its scattering information is like displaying a looped DNA through biochemical effects in that the lawful account that the final information is genuinely information about the object in question amounts to little more than an abductive inference of the presence of the object.

The important points that have emerged in this study of the observability of objects imaged by electron microscope are these. The degree of observability of such objects depends on what it is and the specifics of its imaging. Certainly none of these things, DNA, the microstructure of crystals, or single atoms, is observable to the unassisted human body. It is clear also that the imaging processes of an electron microscope are such that the observations are relatively more indirect (in the modified Shapere sense) than are observations with an optical microscope. In the optical case, the same photons which interact with the specimen also interact with the human observer. The analogous directness is not possible with the electron microscope as humans are not equipped to interpret information by electrons as they can the information carried by groups of photons.

All cases of electron-microscope imaging involve massive support by auxiliary theories which describe the properties of electrons, their scattering and their response to the electron-optical system. But it is not the amount of theoretical support which is distinctive in the case; rather it is the relation between that theoretical support and the theory of the specimen. It is a more reliable claim of information-acquisition to rely on electron-atom scattering theory to probe the shape of a cell than it is to use the same to report on the presence of an atom.

Tectonic plates

As a final example of meaningful information with an apparatus but not with the human body by itself, consider the observability of tectonic plates. The current geophysical description of the earth shows the outer most feature to be a rigid layer, the lithosphere, which is roughly 100 km thick. This layer lies over a more plastic and thicker layer (roughly 300 km thick), the asthenosphere. In the 1960's, geophysicists suggested that the lithosphere is broken into huge plates like a spherical jigsaw puzzle with ten or so pieces that cover the globe. Furthermore, these tectonic plates move about relative to each other and relative to a reference point in the earth's core. The relevant question to ask here is what is involved in observing these plates which are among the most fundamentally important objects studied by geophysics. Being faithful to the schema of assessing observability for an object-property pair requires identifying the specific plate properties that one would expect to observe. I will stick to the easy ones, the plate's shape, that is, the location of its boundaries, and its motion relative to another plate.

Observing that the plate has such and such a shape, that its boundary runs up the east coast of Asia, east to the southern coast of Alaska, down the west coast of North American, etc., would be observation of distinctively plate-tectonic properties. It is reports of this kind of observation that will be analyzed here.

The signals which carry information about a plate to a scientist are seismic waves. These material waves of earth motion transport energy and information in a way closely analogous to energy and information transport by electromagnetic waves or sound waves. So probing the structure of the earth with seismic waves is in many respects similar to probing the structure of a cell using light and a microscope. The heart of the operation is to receive the waves, seismic or light, which have interacted with the specimen and to debrief those waves for any information about the specimen that they gained in interaction.

The seismic signal arrives and is recorded at a station or a network of stations on the earth's surface. The basic information born by the wave is its arrival time at a station, the components of polarization, the amplitude of motion at the station and the frequency. These data must contain the information about the structure of the features of the earth with which the wave has interacted. To reconstruct that information will require laws of information relation from supporting theories.

There are two manners of using seismic waves to probe the boundaries of tectonic plates. One exploits the phenomenon that the moving plates are themselves sources of seismic energy. Analyzing the waves produced by the plates themselves is a process analogous to analyzing a star by the light or neutrinos it emits. The other way to get information about the plate is by studying seismic waves which are produced elsewhere, by a distant earthquake, but which are incident on the plate and through the interaction collect information about the plate. Analyzing these waves is observing the plate much as one observes through a microscope where the light carries information about the specimen in virtue of their interaction. The two different ways of using seismic waves to observe plates require different physical laws to account for the information.

Viewing a tectonic plate as a source of seismic waves is done by conventional seismology. A seismic event is recorded by a few seismic stations and, by comparing arrival times, the focus of the event is located. Simply by mapping the foci of earthquakes as they occur over time, lines of seismic activity emerge. (See, for example, Bott 1982, 132). These lines are interpreted as coinciding with the boundaries of the plates, the reasoning being that the plates in motion generate earthquakes as they slide along or plunge beneath one another.

The waves from plate-generated seismic events offer more information than just the crude outline of the plate. More detail can be filled in by noticing the depth of the events. One finds undeniable pattern in that there are long lines of shallow foci, and elsewhere areas where the foci descend at a 45 degree dip, normal to the line identified as the plate boundary. The shallow foci show the lines of ridges and transform faulting where the plates are diverging and sliding laterally, respectively. The 45 degree dip is identified as a subduction zone where plates are converging and one is bending down to slip under the other. Further information about the subduction zone, more detailed information of the plate boundary, is carried in the amplitude of the waves generated by events in the area. By correlating data on the wave amplitudes as measured at several stations near the event epicenter, one obtains information of the attenuation of the wave along the different paths to the stations and the correlative information about the wave-transmission quality (called simply, Q) of the material between source and station. One finds that the line of earthquakes along the 45 degree dip are in an area of high Q (low attenuation) but surrounded by material of low Q. By associating high Q with a rigid and therefore cold material (rigid materials transport both compressional and shear waves with less attenuation than do soft materials), the Q-information translates into a picture of a tongue of cold, rigid lithosphere plunging down into the warmer, more plastic asthenosphere, a picture of the plate boundary.

The limitations of this method of imaging the shape of a plate are that the data are relatively local and give only a radial picture of the shape. To get a picture of the whole plate one must put together many observations, sticking pins in a map to see the shape. Also, the data of Q values are ray-averaged values in the sense that any attenuation of the wave could have happened anywhere along the ray. Reconstructing structural details from the averaged data calls for either an unhealthy dose of knowing what the picture ought to look like, or using data from many intersecting rays. This latter technique is most effectively done using rays from seismic events that occur not only at the plate boundary but from distant points in the earth as well. The limitations are best overcome, that is, by using seismic waves which interact with the plate but which do not necessarily originate from the plate.

This technique for imaging large features of the earth is known as seismic tomography and has become feasible only in this decade, relying as it does on large, fast computers to assimilate vast amounts of data. The way it works in this: For each station and each event used, the arrival time of the wave is recorded along with information of the focus of the event (from independent, conventional seismological analysis) and information of the trajectory of the ray. Given a general idea of the radial structure of the earth, the thickness of the lithosphere, asthenosphere, etc., one can compute the average velocity of a seismic wave along any particular trajectory. The clues to structure as uncovered by this technique are found in deviations from the expected velocities, velocity anomalies, as they are called. Each station can report a velocity anomaly as occurring only somewhere along the ray, but given many stations all over the earth, many crisscrossing rays are analyzed for velocity anomalies and the sources of anomaly can be pinpointed. The result is a CAT scan of the earth using seismic waves in place of x-rays. The image is a three dimensional map of velocity anomalies, places where seismic waves travel relatively faster or slower.

A velocity anomaly map reveals nothing about the physical structure of the earth or the shape of a plate without some laws of correlation between velocity and constitution, and between constitution and shape. The propagation of a material wave can be studied in the laboratory, leading to the correlation that materials which propagate waves relatively faster are the more rigid, colder materials. Slower wave velocity correlates to more plastic, hotter material. Thus, anomalous fast spots are seen to be rigid, cold, and, by geological inference, old materials. Slow anomalies reveal plastic, warn, young material. There are also velocity anomalies which show a directional preference in the sense that the wave velocity is greater in one direction than in another. This is known as "azimuthal anisotropy" (Anderson and Dziewonski 1984, 65) and is related to a compositional feature by pointing out that much of the outer layers of the earth are composed of olivine crystals which are anisotropic in their propagation of seismic waves. If the crystals are aligned in an area, that area would show azimuthally anisotropic seismic velocities. The crystals could be aligned by flowing material as, for example, by sliding plates of flowing magma in a convection current. This is structural information found nested in velocity anomaly information. The current state of the art for imaging three dimensional structural features of the earth, as reported by Anderson and Dziewonski is resolution of features with horizontal size of 2000 km and vertical extent of 100 km. These pictures are the result of analyzing information from 500,000 rays.

From the velocity anomaly information, a three dimensional feature map is produced. At a level of 100 km below the surface, slow spots (that is, areas of relatively slow wave velocity) are seen at the diverging and converging boundaries of the plates. These are interpreted as reports of hot material, rising in the convection current which moves the plates in the diverging zones, and from hot material displaced upward by subduction in the converging zones. Between these zones, in the middle of the plate, the velocity is fast, indicating cold, older material.

At 400 km depth the velocity map is fast at the converging, subduction zones, indicating the cold plate being forced down. At this depth which is under the lithosphere and hence under the main body of the plate, the velocity is slow under the middle of the plate. There is also azimuthal anisotropy here, showing the waves moving faster in the horizontal direction than in the vertical. Near the convergent zones the anisotropy is reversed, being faster in the vertical direction than in the horizontal. Correlating the generally slow speed with hot material and associating the faster direction of anisotropic velocity with flowing material, the velocity anomaly information reports a lateral flow of hot material from subduction zone across to divergent zone, a process which looks like a shallow convection cell. These velocity anomaly data are shown in figure 11, together with the correlated constitutional and structural information in parentheses and the picture of the plate as constructed from the structural information. As a vertical cross section, figure 11 hides the abilities of seismic tomography to display the three dimensional structure. For a better idea of what the image can look like, see Anderson and Dziewonski (1984, 63, 64 and 66).

The fundamentals of observing a tectonic plate then involve interaction between seismic waves and features of the plate. Different parts of the plate can either speed the waves up or slow them down, giving the waves structural information which they carry to the seismological stations and report in terms of arrival times.

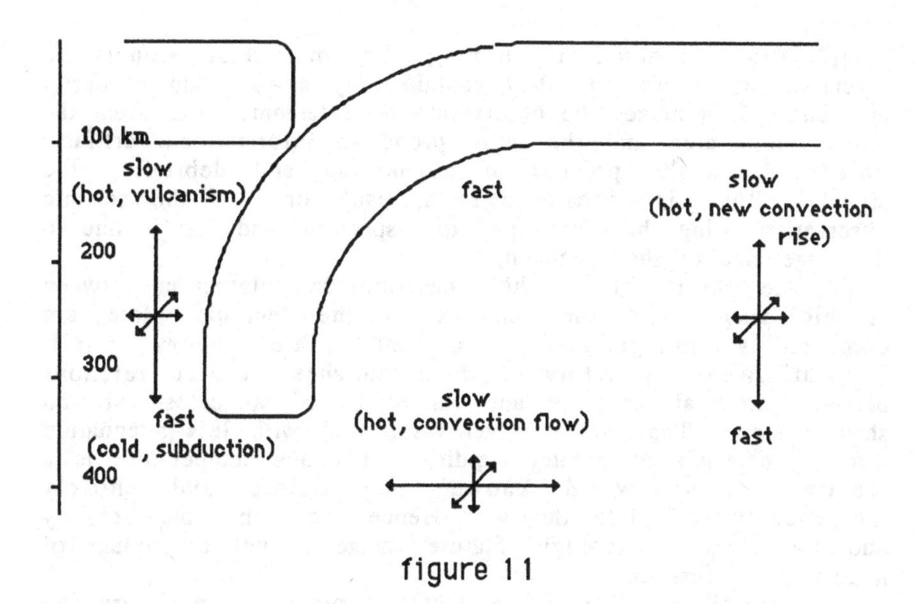

figure 11

It is interesting to note some similarities between the seismic observing and observing with an electron microscope, particularly a scanning electron microscope. In both cases the final image is the product of electronic manipulation of the information from the waves which probe the specimen to form an image of the specimen. Since the human body is unable to process the information carried by seismic waves or electrons, this information must be translated to photons, an informational signal we can use. The final image in both cases presents features which are easily noticeable. (This is believable for the seismic case only by looking at the images in Anderson and Dzeiwonski. Figure 11 above is just a pedagogical sketch to show how the information is correlated, and it is not intended as the final picture.) But this noticeability is true only at the final stage of imaging, and a glance at the information signal at any stage closer to the object would incomprehensible. This is true of both seismic tomography and electron microscopy. But in the latter case the intermediate stages of information are relatively inaccessible. You couldn't glance at the intermediate stages even if you wanted to, since the whole process is contained in a neat instrument which shows only the information in final form. Contrast this with the situation of seismic tomography in which the information is spread out all over the world. The complication involved in mixing the individual signals to assemble the information they contain only as a group is easily apparent. This makes the observation of a tectonic plate seem the more complicated and the more prone to informational artifacts introduced in the process of assembling and debriefing the signal. But this appearance is a result only of the seismic observation being the more open to inspection, and this is due to the large size of the specimen.

The physical laws which describe the interaction between seismic waves and the features of the tectonic plate are comfortably independent of the plate theory itself. Straightforward laboratory work establishes general relations between material structure and its effect of compressional and shear waves. That waves travel faster and with less attenuation through materials of greater rigidity and cooler temperature is a feature of the world known long before and entirely independently of plate theory. Hence the step from velocity anomaly image to geologic feature image is not a product of nepotistic inference.

The observability of a tectonic plate is an interesting example to study because the plate is both a key object in the world of geophysics, and one of the many vague cases of observability which van Fraassen indicates are inevitable and harmless. It is unclear whether a tectonic plate is the kind of thing such that there is someplace one can go and observe it with one's own eyes. Surely it is not too small, but maybe it is too large. If one is in a position to see a section of a plate boundary, to see the creep line of the San Andreas fault, for example, then one is too close to see the entire boundary. And at a distance great enough that the entire extent of a plate would be visible, the details of the boundaries would be too fine to distinguish. In a case like this it is important to be clear about what property it is of the plate that one might observe. It might be van Fraassen-observable, for example, that there is a discontinuity in the lithosphere, and edge of a plate in Western California, or that the relative motion is roughly 6 cm/yr. But it would not be similarly observable that the plate is shaped as _____, since it is too large to be seen at one time. Properties of the whole plate are not accessible to unaided observation. So by van Fraassen's criterion, parts of a plate may be observable while the whole plate is not.

The plate tectonic model is too important a part of geophysics to be left in this ambiguous state of observability. The vagueness is resolved, I claim, by evaluating the observing of plates for what it is in terms of acquisition of information rather than trying to assign it a simple observable-unobservable status.

By following the flow of information in seismic imaging one sees what it means to seismically observe a plate, that is, what information about the plate is relayed and how it is nested and must be gleaned from the basic signal. That is the fuller understanding of the observability of tectonic plates.

3. PERCEIVABLE THINGS

The examples of things unperceivable in fact have indicated ways in which entities can interact with observing apparatus such that the apparatus can process object-information into an informational form accessible to humans. The examples in this section on perceivable things are more immediately observable in the sense that the information of <x, P> is already in a form accessible to human perception and no machine is necessary to mediate the observation. Such things which can interact immediately with human sense organs in a way that information is conveyed, I refer to as being perceivable. It is still relevant in these cases to investigate the nature of the inference, if there is any, involved in relating the human body-state information to the object-state information. It is also still appropriate to evaluate the directness of the observation by detailing the transport of information from object to observer.

Heat

To be clear about the observability of heat, one must first be clear on what heat is, or, more exactly, how the term "heat" is to be used. I will be discussing the standard thermodynamic sense in which "heat" refers to energy as it is transferred from one place to another. Heat in this sense is a process which can occur in one of three ways, conduction, convection, or radiation. Heat is not a state of an object or of a system, and this is where the strict thermodynamic concept differs from the more common notion of heat. Outside of the thermodynamic context heat is often regarded as a state as well as a process, and static systems are said to manifest heat, or event to have heat. But the thermodynamic sense which I will use here, carefully distinguishes between the state variable of internal energy U, and the process variable of heat Q. (This Q has nothing to do with the seismic quality Q). Where there is no change in the state of a system there is no heat. Hot things, as measured by a thermometer, do not necessarily involve heat. It is only if thermal energy is being given up to or accepted from the surroundings that heat is involved.

This specifies the entity of which to investigate the observability. It also points out that "heat" as used in thermodynamics is a theoretical term. The meaning of the term is specified by its role in the theory and by its association with other theoretical terms like "internal energy" and "work". And like "work", the theoretical meaning of "heat", differs from the general meaning of the English word. Work, as heat, in the theoretical sense refers to a process and not a state. Work is the flow of mechanical energy, whereas heat is the transfer of thermal, random energy, and it is the observability of the process to which this latter theoretical term refers which is the topic here. It will turn out that the theoretical term "heat" refers to an entity which is observable in a quite immediate and direct way, thus incidentally underscoring the independence between the theoretical nature of a term and the observability of the entity to which it refers.

The focus is on observing the process of transfer of thermal energy, heat. Instead of asking whether or not heat is observable, ask the more open question of what it means to say that heat is observable. So rather than fitting heat into a category of observability, we are understanding the nature of observability in part in terms of the reception of information about heat.

The observability of heat has implications for the empirical status of the first law of thermodynamics. That law is the statement of conversation of energy in thermodynamic processes, stating that the change in internal energy ΔU of a system must be accounted for as heat Q in or out of the system, and/or work W done on or by the system. Mathematically rendered:

$$\Delta U = Q + W$$

The work W is easily observable since it involves macroscopic motion. Consider the standard heuristic example of a thermodynamic system, a cylinder of gas with a moveable piston, the motion of which changes the volume of the gas (figure 12). In this case the work done on the system is equal to the pressure P of the gas times the change in volume ΔV (that is, $W = -P \cdot \Delta V$, where the minus sign is an artifact of the choice of positive work representing work *on*, not *by*, the system). Change in the volume is manifested by motion of the piston and indicates that work is being done.

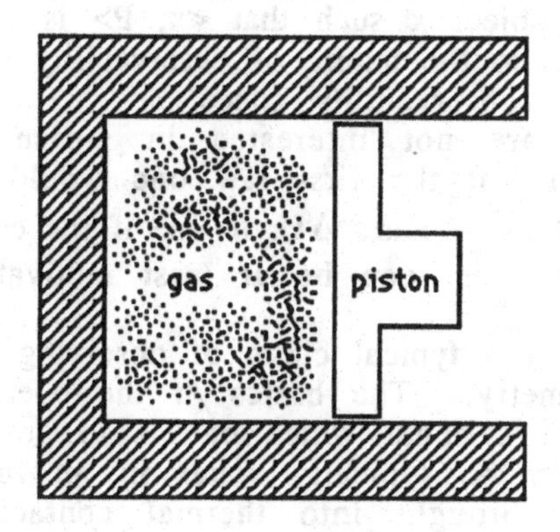

figure 12

Change in internal energy is also amenable to observation and is manifested as change in the temperature of the gas. (This statement is an oversimplification in that it assumes that the gas is ideal and so ignores rotational and vibrational aspects of internal energy. But the place to sort out the implications of this sloppiness would be in a case study of the observability of internal energy.) Observing ΔU is done with the aid of a thermometer, by observing initial and final temperatures and inferring the ΔU-information nested in ΔT-information. The question of the observational status of the first law then hinges on the observability of the thermal energy flow process Q.

Now is the time to distinguish between observing and measuring. Measuring involves quantity, assigning numbers to features of the world. Measurement would require information of the form "x is P to degree n" and insofar as this information can be obtained from x itself, measuring is a subclass of observing. But observation in general can be accomplished entirely without numbering or even ordering. One might observe, for example, that the fire is radiating, where the information is without magnitude. And it is to this nonquantitative kind of observing that I focus the treatment of heat. In the object-property ordered pair schema, the question is of the observability of <x, P> where x some macroscopic object and P is the property of exporting of importing heat. Then heat can be classified as perceivable in the derivative sense insofar as there exists some object x such that <x, P> is perceivable. The question remains as to the nature of observability of <x, P>, that is, its location in the observability space.

Hence we are not interested in measuring Q but only observing Q. To test the first law one would have to actually measure Q, as well as W and ΔU, to see if the equality holds. But acceptance of the claim can be at least motivated by qualitative observation.

Consider first a typical claim of observing Q, namely in the practice of calorimetry. The basics of the operation involve two systems, the object x about which one is to get heat information, and another system A, typically water in an insulated container, with which x is brought into thermal contact. There is no exchange of material between x and A, and no work done on or by x. The two systems, once in contact, are allowed to come to thermal equilibrium, meaning that they are at the same temperature. Then by noting a difference between the initial (before introducing x) temperature of A and the final (at thermal equilibrium with x) temperature of A, that is, by noticing that macroscopic correlate to change in temperature, namely change in length of mercury column in a thermometer, one can claim to be getting information of heat flowing to or from x. Alternatively, if the A-system manifests a change of phase, the liquid water freezes, say, or boils, then from this observation one can infer a heat process even without noting temperature. These simple observations are qualitative and reveal only that energy has been transferred but do not indicate how much energy has been traded. The observation is easily upgraded to a measurement by introducing heat capacity or latent heat of fusion or vaporization, as characteristic properties of the substance in A.

While these procedures indicate an exchange of energy they do so by presenting only initial and final *state* information. And as heat is essentially a process, there is a sense in which observing end-point states misses something important. One cannot reasonably claim to have watched a baseball game by noting the blank state of the scoreboard in the morning and its evening condition of being lit with numbers, even if it is the case that the scoreboard changes only during a game. To watch the game, like observing heat, requires observation of the event. In the case of heat, this would mean observation which is sensitive to one of the three heat processes, conduction, convection, or radiation. Take conduction. The general description of heat flow by conduction is:

$$Q = kA\left(\frac{\Delta T}{\Delta x}\right)\Delta t$$

where t is elapsed time, A is cross sectional area, k is a constant of conductivity characteristic to the material, and $\Delta T/\Delta x$ is the change in temperature per distance, the temperature gradient. Given temperature gradient information one can infer conduction information, that is information of the heat process in progress. Comparing temperature observations, again with the aid of a thermometer, at spatially separated points leads to information of Q.

Both the conduction-observing process and the calorimetry described above include the use of observing apparatus to observe Q. The heat information is herded into a thermometer and therein processed to be visually accessible to the scientist. Heat can be observed with this assistance, but it can also be observed more immediately with no tools at all. The body, that is, can itself serve as calorimeter and conduction detecting device. It may be relatively imprecise in the sense that direct-body results may not be repeatable, transitive, or in agreement with instrument results, but that is more a problem for the measurement of heat than for the simpler observation of the form, that x produces heat.

The claim is that heat is observable in the sense of not requiring an intermediate apparatus to process information. An object can interact with the human body and produce a body-state which contains the information that the object is exporting (or importing) heat. Heat, in other words, is perceivable. It is also, if the claim is true, observable by van Fraassen's standards since there are circumstances under which one can see (feel) for one's self, that is, without mechanical assistance, the heat process. One wants now to look carefully at the interaction between object and body and to trace the conveyance of information about heat, both to understand and to verify the claim that heat is simply observable.

Consider the following simple setup in which the body is used as a kind of calorimeter, sensitive to the flow of thermal energy to or from an object x. x is thermally isolated from the world behind adiabatic walls, except for a small hole which allows contact with a part of a human observer (figure 13).

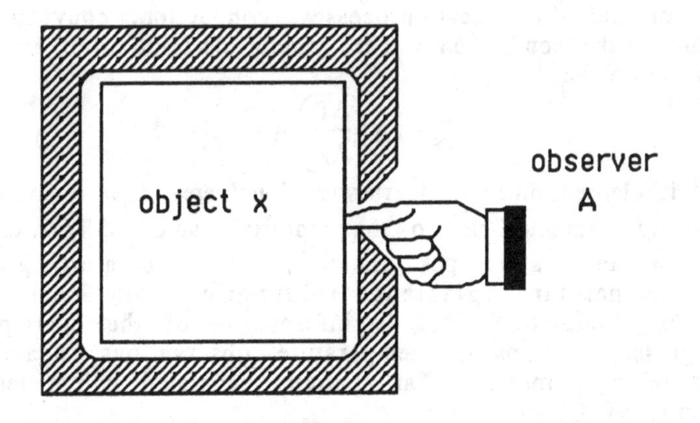

figure 13

The question being asked is of information of the heat process of x, not the observer A, though they are easily related as $Q_x = -Q_A$, there being no work done. We want to observe the object-property pair <x, P> where P is the property of exporting (or importing) heat. To see how such an observation is possible we turn, as is policy, to the physics and physiology of the situation.

The physiological data are typically presented in terms of the correlation between the object temperature T_x and signals from nerves associated with thermoreceptors near the skin (Wyburn 1964, and Hensel 1973). There are two kinds of thermoreceptors, specialized to report sensations as hot or cold. The following results are reported by physiological study. A sensible impulse is sent if the temperature at a receptor (and by approximation, the temperature of x, T_x) is changing in time, that is if $T_x \neq 0$. This response happens at any temperature T_x. There is also signalling from receptors even if the temperature is constant as long as T_x is significantly different from normal skin temperature, specifically, if $T_x < 20$ C, or $T_x > 40$ C. The impulses from the receptors are independent of the temperature gradient in the direction perpendicular to the surface of the skin. That is, there are cases in which there is a nonzero gradient along the skin-depth axis in the neighborhood of a receptor, with no signal sent from the receptor (Hensel 1973, 318).

So the feeling of hot or cold depends on temperature T_x and its change in time T_x. It is the fact that the change of temperature is an aspect of sensation that makes this an experience which includes information of the process of heat and not simply of the state of internal energy or temperature. From the data presented, the first law of thermodynamics, the fact that the skin itself is not a source of heat, and the law of heat conduction (or the second law of thermodynamics), it can be shown that sensation reports from thermoreceptors are indicative of heat processes of the object x. Heat information is nested in information of the thermo-sense organs. It will be of interest to fully expose the inference in this case to make it a clear example of the orthogonality of directness (it is quite direct and immediate) and amount of interpretation (there is a lot). To this end we can display the informational unpacking step by step in a formal deduction. For the argument, let I = "there is a sensible impulse from a hot or cold receptor", τ = "$T_x \neq 0$", B = the fact that 20 C < T_x < 40 C, and Q = "there is heat process associated with object x":

1. $(-B \vee \tau) \leftrightarrow I$	empirical data
2. $\tau \rightarrow Q$	first law, given that W = 0
3. $B \,\&\, I \rightarrow \tau$	1
4. $B \,\&\, I \rightarrow Q$	1, 2
5. $-B \rightarrow Tx = T_a$	since $T_a = T_{body}$ = 37 C
6. $T_x \neq T_a \leftrightarrow Q$	law of conduction
7. $-B \rightarrow Q$	5, 6
8. $I \rightarrow Q$	4, 7

Hence, if there is an impulse, there is heat and by the informational account of observation, the property of heat can be observed. It can also be shown from the fact that spatial changes in temperature do not necessarily produce impulses from the receptors, that heat can go undetected. It is *not* the case that $Q \rightarrow I$.

If it feels hot or cold, one is licensed to say that there is heat coming or going, respectively, from or to the object x. So one can be informed not only that heat is happening but also the direction in which energy is moving. In general cases of touching an object and having a sensation of hot or cold, one gets information of the heat flow between the object and one's body, but not of the total heat process of the object. The object could be transferring energy to or from other parts of its environment at the same time, that is, unless it is within the kind of adiabatic container as in figure 13. Another noteworthy aspect of the heat observation process as described here is that the act of observation facilitates the process being observed. Thermal energy moves from object to finger only when there is a difference in the temperature between finger and object. By touching it, the observer brings about the heat process in the object. It is also true that this is the only kind of heat event that can be observed immediately. One can observe without the assistance of some observing apparatus only the energy-flow process from x to A in which A is a part of the observer's body. Heat flow from the window to the cold air outside cannot be felt with the bodily thermoreceptors. It is only for objects in thermal communication with the body that one can claim to immediately observe their heat properties, and with these it is only the information or heat to or from the body itself which is available. And one other restriction on the kind of objects for which heat is observable to the unaided human body is that they must be at a temperature which is not drastically (many orders of magnitude) different from the temperature of the human body. Transportation difficulties aside, one would still be unable to collect heat conduction information of the interior of the sun with the tip of one's finger.

The conclusions about observing heat can be summarized by locating the property of exporting (or importing) heat in the observability space. It is, I have argued, perceivable in the sense that there are objects x which can interact with the human body in such a way that the body is put in a state which contains information of the form, "that x is giving (taking) heat". No extra apparatus is necessary to observe heat.

It is interesting to note the restricted class of objects of which one can immediately observe heat processes, the most important restriction being that the event must include the body as a participant. Contrast this with observing work. With the piston-cylinder arrangement as in figure 12, one can observe the motion of the piston and hence observe the work done on or by the system even though the observer's body is not involved in the work being done. The observer neither does work on the system nor is it the recipient of work done by the system. Work can be observed by an innocent bystander. Heat demands participation, and this is a telling restriction on its observability. If this were the only means of observing heat, that is, if it could not also be done with the aid of apparatus such as calorimeters, then thermodynamics might be cast in more anthropocentric terms. There could be, for example, a first law for bodily contract but not for systems in general.

The amount of interpretation to find the heat information nested in the thermoreceptors' signals is evident in the argument above. The important thing to notice though is not so much how many steps there are in the inference but what physical laws are invoked to justify the steps. The conservation of energy, that is, the first law of thermodynamics, and the law of conduction are used to find heat information in sensation. (The second law of thermodynamics can be used in place of the law of conduction.) This indicates that one must know about heat in order to know that it is heat one is observing. There is a close dependence between the physical laws which account for the nomic nesting of information, and the theory of the property being observed. This sort of nepotism disqualifies the bodily observation of heat as a method for testing or even motivating the first law of thermodynamics, for there is no trial of the law of conservation of energy where the conservation of energy must be invoked.

Acceleration

The acceleration of an object is a paradigm of an observable property. The entities of mechanics can be divided into two categories, kinematic and dynamic, corresponding to the description of motion and the cause of motion, respectively. The dynamical entities like forces and mass are often regarded as problematic with respect to observability, but the observability of kinematical properties of a macroscopic object, its position, velocity, and acceleration are rarely questioned. Motion is observable; it is the cause of motion which has uncertain empirical status. And when the empirical status of Newton's second law, $F = ma$, is questioned it is with regard to force and mass but not acceleration. But any thorough evaluation of the empirical standing of the second law should also account for the nature of the observability of acceleration.

Kinematical features are also pivotal in relativity theories in the sense that one distinguishes between relative and absolute motion. Empirically based decisions between competing formulations of mechanics could be based on the difference in observability of relative and absolute motions. Newton's spinning bucket experiment was advertised as a demonstration of both relative acceleration of the water, relative to the bucket, and absolute acceleration of the water. It proves, if nothing else, that the observability of acceleration warrants careful investigation. It also indicates that the observability status of acceleration will have implications for the issue of observability of absolute space. The nature of observability of acceleration will also be relevant to the empirical status of the equivalence principle of general relativity in its formulation that the effects of gravity are empirically indistinguishable from a uniform acceleration of reference frame. The observability of acceleration, in other words, is worth knowing about.

Specialize the discussion to examples appropriate to the above issues, namely Newton's second law of motion and features of relativity theories. To this end, invoke the object-property pair schema by discussing the observability of <x, P> where x is some midsized, solid object like a tennis ball or a brick, something which is unproblematically observable in the sense that the information that x is present is immediate and uncomplicated. Choose the property P to be simply "is accelerating," that is changing speed, direction of motion, or both. This is to be qualitative observation where one is content with information of speeding up, slowing down, turning, etc., and I will not be concerned with our ability to measure acceleration. We will not be looking for information of the form that x is accelerating at a rate of 9.8 m/s^2. This would be another, more complicated issue.

The discussion of acceleration presented here is intended to cover the standard examples of classical mechanics as used in the textbooks and student laboratory demonstrations of Newton's laws of motion. What is said here should apply straightforwardly to Atwood's machines, objects in free-fall at the earth's surface, simple pendula, and masses on springs.

A rigorous application of the interaction-information account of observability to this case involves three tasks. One wants first to know how the information is brought to an organ of sensation. This involves the medium of information and the interaction between object and medium, and between medium and body. Then there is the resulting state of the observer, the A-state as it is called in chapter two, to be described. And finally one must account for the information by finding the object S-state information which is nested in the A-state information. The issue of the observability of acceleration (and motion in general) is particularly well suited to an empirically based study as is being done here. It is a good issue to put to the relevant sciences, the physics of motion and the physiology of vision, because the relevant question can be put very cleanly. Do we perceive motion itself, or do we perceive only position and infer from change in position, relying on memory, motion? The relevant sciences have an answer to this question, but certain important aspects of the answer are incomplete. For this reason the observability of acceleration is interesting to study as an unfinished research program.

Here are the facts: Consider the example of observing a single object in motion with respect to a background such as a wall or a tabletop. The medium of information is a group of photons which are reflected from the object and from the background and focused by the lens of the eye to illuminate the retina. The focusing is such as to present the retina with a field of contrasting light and dark which is an image of the actual object-background field. So an object in motion is imaged as a changing contrast field on the retina with a relatively brighter (or lighter) patch in motion over the retina. This is the chain of interaction between object in motion and sense organ.

The informational state which results from this interaction is described in terms of the signals from nerve cells which are part of the visual system. (I draw on information from Sekular, 1975.) Measurements comparing nerve impulses with visual stimulus indicate that the human visual system includes not only detectors of position but also a variety of motion detectors. These are cells in the visual system which signal when and only when the image focused on the retina involves motion. There are, for example, directional detectors. These are cells which send signals if and only if the stimulus moves along some particular axis, up or down, left or right, or at some specific angle. A horizontal motion detector, for example, would signal if some aspect of the image field moved horizontally. It would not fire if the image was motionless or if it moved vertically or at some appreciable angle with respect to the horizontal. There are also speed detectors in the form of separate impulse channels for separate, though overlapping, ranges of speed.

The motion detectors are further specialized in that some are sensitive to motion of the entire retinal image while others are sensitive only to motion within the frame. The former would signal if, for example, the object and background had no relative motion but the observer moved her head, thereby giving the whole view a movement. The relative motion detectors would signal in the event of the object moving across the background, that is, relative motion in the stimulus.

It is unimportant for our purposes exactly where in the visual system the motion detectors are located. Whether they are part of the eye itself, or part of the brain, or somewhere in between, does not matter as long as they are somewhere in the hardware of the system. Their existence indicates that one does not infer motion information from observations of position. Rather, velocity is perceived as directly (in the sense of not requiring inference) as is position, and information of the form "is moving straight down" is on observational par with "is just left of center" or even "is red".

Reports of motion detectors describe signals for direction and speed but there is no mention of acceleration detectors. These would be cells in the visual system which signal in the event of change of speed or change of direction, just as speed and direction detectors signal change in position. There is no denial that acceleration detectors exist, only no word that they do. Having found speed and direction detectors, the research seems to have rested, and it is here that I locate the unfinished business of the scientific treatment of the observability of acceleration. If there were acceleration detectors which functioned analogously to direction and speed detectors, the investigation would be finished and acceleration would be as simply perceivable as position or color. The search for acceleration detectors then would be a fruitful research project for the empirical study of observability.

In the absence of evidence on acceleration detectors one can continue to investigate the acquisition of information by more complicated, indirect experimental means. There is evidence from study of stimulus and response that acceleration is directly perceived in the sense that no inference of combining other, more immediate data is required. Instead of physiologically probing for nerve signals, Rosenbaum (1975) investigated the perception of acceleration by studying the overt behavior of people in response to controlled acceleration stimulus. His data result in a comparison between the "objective acceleration" of the stimulus and the perceiver's "subjective acceleration" as manifested in their behavior. The experiment is a test of the accuracy of one's perception of acceleration and Rosenbaum uses the results to draw a conclusion about the immediacy of the acceleration percept. He evaluates the data through two a priori assumptions. For one, if acceleration information is inferred from more direct information of position and time, then stimuli presented to the observer over longer spatial distance and longer temporal duration will allow more accurate evaluation of acceleration. In other words, if there are no acceleration detectors and acceleration information must be put together by noting change in motion or change in change in position, then the more data on position one is given, the longer one gets to look at the object, the more accurate will be one's assessment of acceleration. The second of Rosenbaum's working hypotheses is that if acceleration information is inferred from more basic data then forewarning the observer that the object will be accelerating will enhance the accuracy of perceived acceleration. The assumption is that if one is prepared to make the inference then one will be able to do so with more reliability than if one is surprised into the task.

Rosenbaum's results (1975, 400) are that the accuracy of perceiving acceleration is independent of the time and distance of exposure to the stimulus. It is also insensitive to advance knowledge, as two groups of observers, one forewarned of acceleration, one surprised by it, performed nearly identically. The conclusion is that there is no inference involved in the perception of acceleration. One perceives acceleration as directly (noninferentially) as one perceives velocity (of which Rosenbaum has done similar experiments). Rosenbaum even gives the perception of acceleration a primacy over velocity, claiming that the perceptual system is, "primarily tuned to acceleration rather than to constant velocity" (1975, 402). He draws this conclusion on the basis of a premise that the perceptual system is more responsive to change than stasis, and acceleration shows more change than does constant velocity. I find this final step in Rosenbaum's argument dubious because it invites another step and does not indicate the end of a potentially endless inflation of change. Why settle for acceleration when you can have change in acceleration, or better, change in change of acceleration?

Rosenbaum's results are encouraging for the noninferential perception of acceleration, if not for the hierarchy of acceleration being the primary stimulus of the perception of motion. But the results lack the clear signal of direct acceleration perception as would the physiological location of acceleration detectors. His data are only the outward signs of acceleration detectors in the perceptual system, and in the absence of the clear signal it is prudent to consider the possibility that there are not detectors of acceleration. It is worth making sure that there is acceleration information contained in the signals of position, speed, and direction.

Note first that neither signals of position nor signals of speed contain acceleration information individually. Nor does a coincident pair of signals, position and speed at one instant, inform of acceleration. (I am suppressing the directional dimension here only to simplify the argument. It could be included with no difficult or important alteration.) If there is no direct detector of acceleration, then the perceiving system must mix information from signals of position and speed at two different times. One must also have access to the information that these signals are reports from different times, that is, there must be information of succession or at least of noncoincidence. Then from noncoincident information of position z and noncoincident information of speed v, there is information of change in speed and change in position, that is, $\Delta v/\Delta z$. Given that the speed is related to position and time as $v = \Delta z/\Delta t$, the acceleration information $a = \Delta v/\Delta t$ is nomically nested in the change-in-speed per change-in-position information as $a = \Delta v/\Delta t = v \cdot (\Delta v/\Delta z)$. Thus is acceleration information to be found in the signals of position and speed detectors.

There are two interesting and important points to be seen in this analysis of observing acceleration. One is that being equipped, as physiological evidence indicates we are, with position and velocity detectors leads first to the information of $\Delta v/\Delta z$, from which the acceleration information $\Delta v/\Delta t$ must be unpacked. The immediate change-in-velocity information is presented as change with respect to space rather than time. This suggests that a change of velocity in space would be more noticeable than a change in time. And this underscores Galileo's achievement in redescribing the acceleration of a freely falling object in terms of speed change in time rather than distance fallen. He overcame not only a firm tradition in the science of mechanics but also the greater noticeability of $\Delta v/\Delta z$ over $\Delta v/\Delta t$, which was likely responsible for the tradition. (Hanson 1958, Chapter 2, emphasizes this aspect of Galileo's achievement.)

The other interesting point regarding the observation of acceleration using position and velocity detectors is that it does not call on any perception of time. One does not need access to metrical temporal information to get qualitative acceleration information. One needs only information of noncoincidence or order (to distinguish speeding up from slowing down). The interesting thing is that this is not true if the observer does not have motion detectors but has only position detectors. To see why this is the case, consider an aside on a typical laboratory project for physics students to observe and measure acceleration. The setup is pictured in figure 14a.

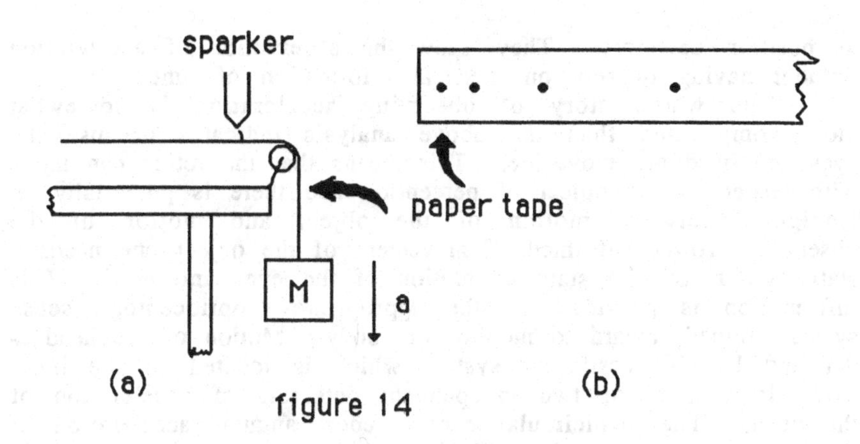

figure 14

As M accelerates it draws a tape of waxed paper through the sparking apparatus which sparks with some constant, known frequency f, and thereby puts a dot on the paper every $\Delta t = 1/f$ seconds. The data in which acceleration information is to be found consist of the paper tape and its pattern of dots, as in figure 14b. This is relevant to the question of position and motion detectors because the paper tape is a position detector but not a motion detector. One can observe the position z of the dots and their separation Δz and even the change in their separation $\Delta(\Delta z)$ along the length of the tape. This information, together with information of the temporal order of the dots, and the elapsed time between dots, allows the conclusion that the object has accelerated, and a calculation of the magnitude of acceleration. Typically this is done graphically by plotting $(\Delta z/\Delta t)$ versus elapsed time t. The slope of the line indicates acceleration. Noting the temporal succession of dots but without including the metrical temporal information about the dots is not enough to license a conclusion about acceleration. The pattern of dots as in figure 14b could have been produced by drawing the tape at a constant speed through a sparker which marks the tape at steadily increasing intervals. One needs to know at least that the spark frequency is constant in order to get even the qualitative information that the object is accelerating, let alone the rate at which it is accelerating.

The point is that it is an important physiological fact that the human visual system has speed and direction detectors as well as position detectors. They allow the observation of acceleration without having to rely on metrical information of time.

The whole story of observing acceleration is somewhat more complicated than the above analysis indicates because the eyes and head are moveable. This means that the retina can move with respect to an object of perception and there is potentially an ambiguity between motion of the object and motion of the observer. To be informed of movement of the object one needs a status report on the state of motion of the eyes and head. This information is provided by the appropriate proprioceptors, sense systems turned inward to monitor the body. Motion of the head is detected by the vestibular system which is located in the inner ear. It consists of two independent detectors of acceleration of the head. The semicircular canals report angular acceleration of the head about any axis. They serve, in a sense, as a built-in Newton's bucket device and signal when there is relative motion between the containing canals and the inertial fluid inside. The other part of the vestibular system is the utricle which responds to linear acceleration of the head. It is more sensitive to horizontal acceleration than to vertical.

The two components of the vestibular system serve to establish an inertial reference frame with respect to which acceleration of external objects are observed. It is interesting to note that the utricle system is insensitive to any difference between effects of acceleration and effects of gravity. It responds identically to a titling of the head with respect to the gravitationally oriented vertical as it does to horizontal acceleration (Howard 1973, 287). This aspect of the human perceptual system is therefore consistent with the equivalence principle.

The analysis of the observability of acceleration is completed with an evaluation of the dimensions of acceleration observability. Without confronting the question of measuring acceleration, asking, that is, only of qualitative observation, it is clear that the object-property pair < x, P = is accelerating > is perceivable. No extra observing apparatus is required to observe that a falling brick is speeding up. The observation is also relatively direct in the modified Shapere sense. It is as direct as any visually observed entity can be in that the information is brought to the eye by the same photons which interacted with the object. It requires a group of photons extended in time as well as in space, since one needs information of the object at at least two different instants.

The amount of inference required to reconstitute acceleration information from the signals of the visual system depends on whether or not that system includes acceleration detectors along with the position, speed, and direction detectors. If it does, then no inference at all is required. If it does not, then a small amount of inference is required to get acceleration information from position and speed data. The case of the observability of acceleration is a case of an empirical observability study in progress.

Assuming that some inference is done, that is, that there are no acceleration detectors, one can ask about the independence between the physical laws used to support the inference, and the theory of acceleration. But "acceleration" is not a theoretical term in the way that "heat" is. "Acceleration" figures into the formulas and definitions of mechanics, but its own meaning is not fixed by mechanics. Kinematics is not so much an explanatory theory as it is simply a descriptive lexicon. Kinematic claims about acceleration do not require evidence for confirmation in the way that claims about electrons or thermodynamic heat do. In this sense then, there is no danger of nepotism in the account. And clearly their is no use of the theory-of-the-object (the brick or whatever) in the account, except perhaps a need to understand its constitution and ability to reflect light. But this has nothing to do with acceleration and so the independence of this account is near the top of the scale.

Bubble chambers

Elementary particles like photons, electrons, and pions, are often the focus of dispute in the issue of observability. It can be argued (van Fraassen 1980, 17) that the particles themselves are unobservable while the tracks they may leave in a bubble chamber, if they are electrically changed, are observable. This suggests an analysis of the issue at two levels, the observability of the particle for which the bubble chamber is used as a medium of information, and the observability of the picture itself as displayed by the bubble chamber. If the track in a bubble chamber is observable in van Fraassen's sense then there is something to be learned about observation and observability by describing the nature of such an observation. In this section then, I will pursue not the information of an elementary particle but of the manifest results displayed by a bubble chamber. The central task will be a specification of the phenomenon, the one to be saved.

In the interaction-information picture, the use of a bubble chamber is described as an observing apparatus which processes information from interaction with some entity into information which is perceptible and meaningful to the observing scientist. The focus here is on the step between machine and person, the observation of the apparatus state. The results will reflect more on the nature of observation than the modality of observability.

The presentation of the picture in a bubble chamber is in the form of lines of tiny bubbles formed in a superheated liquid, typically liquid hydrogen. What deeper physical events and entities these bubbles indicate, that is, the informational account of their formation, is unimportant here. That would introduce the question of observing elementary particles by means of a bubble chamber rather than the more specific issue of observing the bubble chamber results. It would be an interesting question, the observation *with* (rather than *of*) a bubble chamber. It is a complicated reliance on thermodynamics, electrodynamics, special relativity, and atomic theory only to describe the formation of a bubble (Peyrou 1967).

The observation of bubble chamber results does not happen by looking through the window into the chamber to see the strings of bubbles. Individual events are always presented as photographs of the tracks in the chamber. This is not simply a convenience for the scientist to allow leisurely measurement of the tracks and lasting copies of the information. It is a necessity to photograph the bubble chains to make them visible. The problem is that the bubbles grow so fast that the tracks do not last long enough to be visible by the human perceptual system. At typical operating conditions a bubble chamber displays approximately ten bubbles per centimeter. With this density, a line is best formed with bubbles with a diameter of approximately 0.5 mm. And at normal operating conditions the bubbles grow to this size in approximately 2 ms. By the time half a second has elapsed the bubbles would be over 1 cm across and the track would be indistinguishable. The bubbles would be this large except that the chamber is repressurized and recycled for another event usually ten times per second. (The data for this paragraph are drawn from Ballam and Watt 1977, 85, and Peyrou 1967, 48).

The point is that if you look into a working bubble chamber you see no sign of tracks. The way that the tracks are made visible is by photographing them with high-speed film. The chamber is illuminated with a bright flash of light triggered to coincide with a pulse from the particle accelerator, or more selectively triggered to flash only when peripheral particle counters indicate that an interesting event has taken place. The time delay between the event and the subsequent triggering of the flash of light gives the bubbles their 2 ms to grow to optimum size. For individual events then, the data presented by the bubble chamber in a form which is accessible to a human observer are in the form of a photograph. There is this sense in which you cannot see the tracks in the chamber for yourself.

Photographs of individual events are not, however, the usual data which the scientist observers of the bubble chamber. The more common presentation of results if in the form of statistical information assembled from several events. The statistical data are prepared and presented in the following way. Many two-dimensional photographs of individual events are scanned by nonscientists who are trained to notice specific topological features in the tracks. They pick out the events with interesting (to the scientist) or expected vertices or lines, noting number of branches of the vertex, what happens to each branch down the line, etc.. These selected events are then measured and analyzed with the aid of a computer. For this, a three dimensional image must be reconstructed, either by stereo reconstruction from two separate cameras, or, still in the early stage of development, by holography. Information on locations of tracks are put into a computer such that lengths of paths (and particle lifetimes), curvature (momentum), and sometimes ionization (energy) can be calculated and remembered.

These data are recorded for many events and the cumulative data are then displayed on a single picture. A typical bubble chamber products is a histogram such as mass plot or a Dalitz plot. The histogram mass plot (figure 15a) is the result of measuring the invariant mass M_{12} of two resulting particles in, for example, an event which produces three particles. (The invariant mass is defined as $M_{12}^2 = (E_1 + E_2)^2 - (p_1 + p_2)^2$, where E = energy, p = momentum, and the speed of light c is set equal to 1). The computer then plots the number of events $N(M_{12})$ with invariant mass M_{12} as a function of M_{12}. Sharp peaks in the mass plot, so-called resonances, indicate short lived particles created in the original event and which decay (in too short a distance to produce a visible track) into the two particles whose tracks were measured.

A Dalitz plot (figure 15b) reveals similar sorts of information. Here, one axis is M_{12}^2, and the other is M_{23}^2. Each measured event is represented as a dot, placed in the Dalitz plot according to the values of invariant mass. A concentration of dots along a line as in figure 15b indicates a short-lived particle.

Statistical plots like these are the typical objects of what a scientist observes of a bubble chamber. The useful information in these bubble chamber results are essentially statistical and there is no getting closer to the action than the plot itself. There is no resonance to be observed in the photograph of a single event.

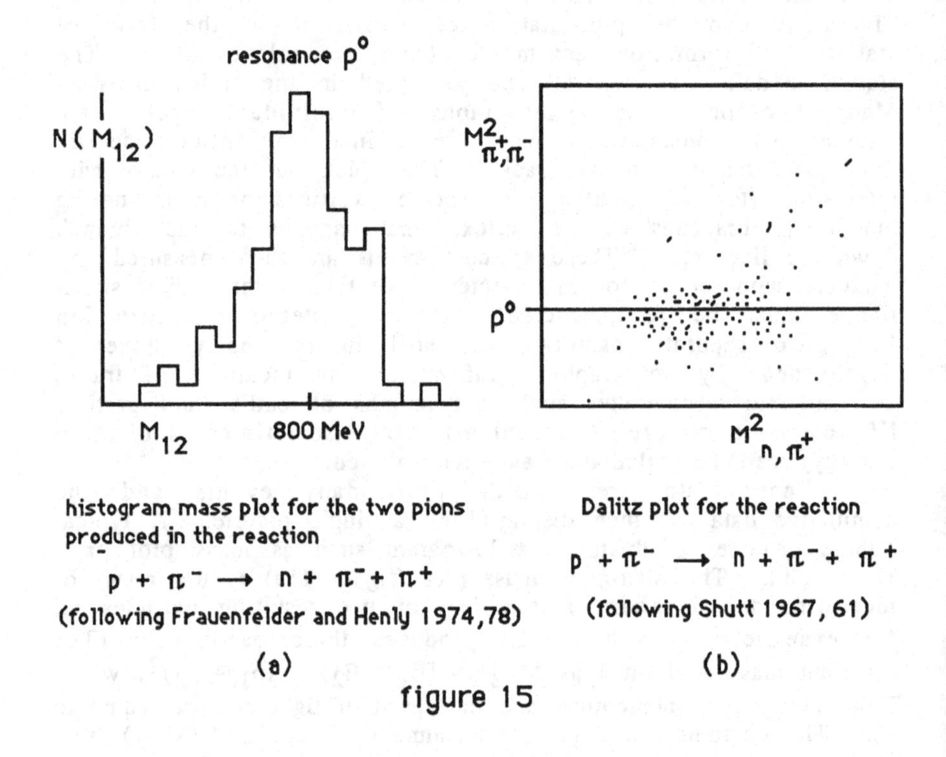

histogram mass plot for the two pions produced in the reaction

$p + \pi^- \longrightarrow n + \pi^- + \pi^+$

(following Frauenfelder and Henly 1974,78)

(a)

Dalitz plot for the reaction

$p + \pi^- \longrightarrow n + \pi^- + \pi^+$

(following Shutt 1967, 61)

(b)

figure 15

What then are the phenomena presented by a bubble chamber? Recall van Fraassen's distinction between the unobservable microparticle and its observable manifestation in the detector. The particle is detected, he says, in a way "based of observation" (1980, 17). The question here is, observation of what, exactly? The results from a bubble chamber can be described in a variety of ways, ranging from theory-neutral sensation reports to thoroughly theory-assisted descriptions of particle interaction. In increasing order of theory dependence, a photograph of a single event can be described as strings of dots....a curved line (or a vertex)....a vapor path in a bubble chamber....a track of a charged particle (or an interaction event with several charged particles)....the track of a π^- (or the event p + $\pi^- \rightarrow n + \pi^- + \pi^+$). A similar sort of progression can be made for the bubble chamber data rendered statistically, the histogram mass plot, for example. It is a steplike graph....a massplot histogram with a bump....a resonance....a short-lived ρ^0. The question is, which entries in these sequences are the phenomena?

Van Fraassen is explicit in disallowing the last entry in either sequence from being regarded as manifest phenomenon. But that leaves several others to choose from, descriptions which differ with respect to the amount of interpretation and background knowledge they require. What counts as the phenomena in this case, depends on the observer, her background knowledge and the use she intends for the observation. For the person who scans the bubble chamber photographs, the phenomenon, the observation, is of curly lines and vertices. At this point there is no regard for the larger context of the photograph, that they are taken at the terminus of a beam of high energy electrons, or that a gap in the pattern will indicate the passage of a neutral particle, for example. For this person, the phenomena are lines and branches. But this in itself is not the kind of phenomenon worth saving. The scanner can *notice* the curly lines and patterns, but noticing, "aha, a curly line" is not observing anything which is relevant to science. The curly-line information must be associated with information of the circumstances of the formation of the curly line.

For the scientist then, the observation which can be useful to the science, the relevant phenomenon, must be of the curly line or statistical graph *plus* the contextual information of the production of the picture. The phenomenon to be saved is of the form, "a curved line of bubbles in a superheated liquid...when a pulse from the accelerator...," or of the form, "a peak in the mass plot at energy .... for events that look like ... which result in circumstances...." One must include the circumstances of the beam, its composition, energy, and orientation with respect to the figures in the photographs. This is not to say that one must be a realist about the beam particles. It does not have to be information that 2 GeV electrons are entering the chamber, but the phenomenon includes at least the information of the outward conditions of the accelerating machine itself.

To give scientific value to the curly lines and steplike graphs one also needs information about the conditions of the bubble chamber. A measured curve is not a phenomenon worth saving without information on the magnetic field where the curve was formed.

The point is that the lines in a photograph or the histogram alone is not a scientific observation, that is, not a phenomenon worth saving, without the contextual information as well. The observation must be such that it can be related to other relevant information such as particle physics theory, and this is only possible given the whole story of the circumstances of the creation of the picture or graph. The scientist does not posit the existence of the unseen ρ^0 particle to explain the observation:

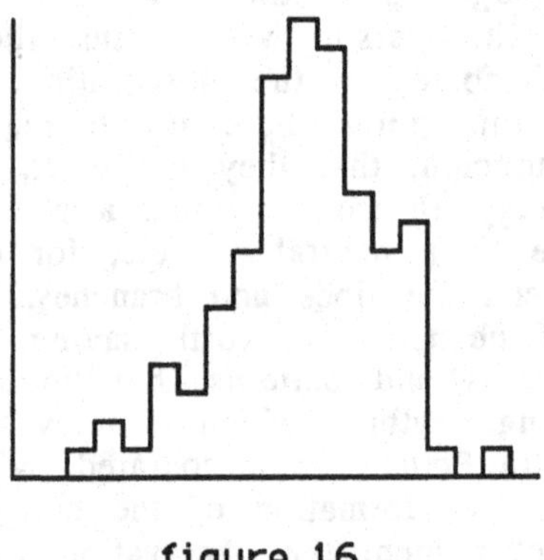

figure 16

but to explain the phenomenon of figure 16 together with the observed circumstances of its happening.

This says something about the theory ladenness of observation. The scanner's observation of "a curly line" requires less theoretical support than does the scientist's observation of "a curly line which results in a bubble chamber at the end of a 1 GeV electron beam...." In the case of the scientific observation, it is possible to insulate the observing process from the theories of particle physics. That is, particle physics may be regarded as inaccessible (a la Fodor) theory in this case. But it is not possible to bar access to the engineering principles of experimental particle physics. One has to know what's going, at least at the settings-of-the-instruments level. The lines in the photographs and the graphs are useful scientific observations only with access to the background information of what happened.

I have postponed introducing the object-property ordered pair schema into the bubble chamber example until now because the important question in this case is of what properties to measure observability. Put a better way, it has taken the presentation of the details of the case to indicate what properties of bubble chamber results function importantly as observations in science. The conclusion of the presentation is that the appropriate application of the <x, P> schema is to regard the object x as some bubble chamber result, either the photograph of a vertex event of a statistical graph such as a histogram mass plot or a Dalitz plot. The property P in the case of a single event must include information of the topology and geometry of the tracks, and the context of creation of the tracks. In the case of the statistical data, the relevant property P must include not only the shape of the picture and the relevance of the axes, but also the circumstances of creation of the events represented.

To observe this kind of object-property pair requires some conceptual preparation on the part of the observer, preparation in the form of knowing of what is going on in the experiment. It is not so much a case of having to interpret the information in the sense of digging out other nested information. It is rather that the information one does get from the photograph of the graph is not useful unless it is combined with other, contextual information. Some of this contextual information is of conceptual origin and some is from observation. One needs a conceptual understanding of the machines being used and of the basics of geometry to make sense of measurement. And one must observe the circumstances of the experiment, the machine parameter settings and relative positioning of apparatus, to get the requisite contextual information.

The case of observing the results of a bubble chamber experiment can be summarized by saying that the observation is not so simple as it casually appears. To say simply that a track in a bubble chamber is observable is deceptive in the complication it hides. One must be clear on what it is that is observed, pointing out the distinction between results presented as photographs of single events, and results presented as statistical plots, the latter typically being the closest the scientist gets to observing bubble chamber results. In either case, the photograph is the closest the human observer can get to observing bubble chamber tracks, they being too short-lived to be observable to the unaided eye.

The most important complication involves identifying the full scope of the observation as it is to be scientifically important. Insofar as the phenomena of a bubble chamber are worth saving, the observations must include the contextual information as well as the information of curly lines.

4. THE ETHERS

The examples presented so far have all dealt with entities of currently accepted sciences. Quarks, photons, heat, and the rest all function in today's scientific description of the world. For the sake of contrast and breadth, it would be worthwhile analyzing the observability of an entity which was part of some scientific description in the past but which has fallen from grace and is now seen as a mistake. There is, of course, a plethora of such things, but the most valuable lesson about observability will come in assessing the important cases, where importance can be judged in terms of the scientific popularity and longevity of a concept. Such things as the various kinds of ethers, luminiferous ether, gravitational ether, caloric etc., and the crystalline spheres of ancient astronomy were essential and widely endorsed aspects of the sciences of their time. Studying the nature of their observability can extend the understanding of the concepts themselves and of observability in general.

Because the analysis of observability is to serve as a preface to a discussion of scientific realism, it is particularly valuable to include the examples of past sciences. Such things as the luminiferous ether and caloric are often used as object lessons by the antirealist (for example, Laudan 1981a). These are concepts which describe unobservable entities, the argument goes, concepts which were successful in the sense of explaining other, observable phenomena, but which, it turns out, describe entities which do not exist. The idea is what with hindsight the ontological status of these things becomes apparent. But the lesson for the realist is not definitive without also clarifying their epistemological status, that is, without carefully clarifying their observability. It has been shown already that the predicate "unobservable" admits degrees and varieties, and the relationship between unobservability of ethereal fluids and their existence is uncertain until the nature of the unobservability is specified.

In discussing ethers and crystalline spheres, entities rejected by today's sciences, there is the implication that mistakes were made, that these theories and their entities should not, given what is now known, have been accepted. The analysis of observability might help to locate mistakes by revealing whether they were associated with the ascription of observability status. This is not to say that the ether theorists made mistaken observations, whatever that might be, but that they might have wrongly evaluated the observability of the entities. Mistakes in observability are certainly possible. Recall that van Fraassen and Foss disagree over the observability of the photon. By some standard then, one of them is mistaken. The question to ask of past theories is whether, by the standards of those theories themselves, observability of the entities was correctly classified. This sort of analysis will indicate whether the mistakes in question were rooted in epistemology, in ascribing observability, or in the ontological conclusions drawn from (or in spite of) the epistemological classification.

A third value in evaluating observability of ethers and the like will be the onotological relevance it suggests for various regions of the observability space. By evaluating the dimensions of observability of the entities of disfavored theories and thereby locating them in observability space, one learns more about the near-neighbors in the space. This proceeds under the assumption of a relation between observability, or some aspect of observability, and ontological credibility. Knowledge of the observability status of some fallen entity, together with the hindsight confidence regarding its ontological status (of crystalline spheres, for example, that they do not exist), suggest that entities of similar observability status deserve similar ontological status. Locating the crystalline spheres in the observability space therefore suggests ontological discredit for its near-neighbors in that space. The point is that if any ontological conclusions are to be drawn from epistemological classifications, the epistemology has to be done carefully, and this includes evaluating the dimensions of observability of the entities of which we have ontological opinions.

To do this, I will focus on one very general and characteristic example of scientific entities which have fallen from grace, the ether. As suggested above, there are many kinds of ethers, different for different phenomena. There have been ethers to explain the mechanisms behind gravity, heat (caloric fluid), light (luminiferous ether), electricity and magnetism. In some theories, a single ethereal substance is invoked to explain all these phenomena under a unified ether theory. But more successful are the ethers which are specific to some action such as light or heat. And even to account for a particular phenomenon such as light there is a creative variety of ethers which are proposed by different scientists to explain the propagation of the action. Ethers can be continuous or particulate fluids, they can be solids, some are homogeneous, some not. There are plenty of crazy models also, of foams and jellies, or even cubicle boxes holding gyroscopes and held together with strips of cloth. (These are reported, though not endorsed, in Duhem 1906, 82-84). For all these models though, some common, essentially ethereal features can be distilled. Following Cantor and Hodges (1981, 2), all ethers are extended throughout space and time (they are ubiquitous), can pass in and out of regular substances (they are subtle), are not preceiveable (and in particular have no perceivable weight, that is, they are imponderable), and can transmit actions and information through space and time. This last property, the transmission of information, is a result of the fact that, "changes in its distribution or its state can cause observable changes in ordinary bodies" (Cantor and Hodge 1981, 2). Taken together, the last two properties, being imperceptable but capable of transmitting actions, indicate that the ether itself is unobservable but its effects are not. Quoting a twentieth century advocate of the ether, "It is commonly said that we have no sense organ for appreciation of the Ether; and we have not any means of appreciating it directly, but we are very much accustomed to appreciate the phenomena which go on it it, or in other words to apprehend its modification" (Lodge 1925, 26). The task here is to understand both aspects of this claim.

There is one complication in presenting the ether (or, more accurately, the ethers) as a scientifically discreditted entity, and that is that even today the ether is not entirely dead. After the special theory of relativity seemed to have banished the ether as unnecessary, the general theory of relativity seems to have revived the idea under the name of curved spacetime. Einstein admitted as much, "According to the general theory of relativity space without ether is unthinkable" (quoted in Cantor and Hodge 1981, 54). Furthermore, the vacuum as described by quantum field theory is an active medium full of short-lived and imperceptable particles and in this sense takes over the role of the ether. So the case of the ontological status of the ether is not closed. But I suspect that there is probably no entity worth considering whose ontological epitaph is fully written. And as the ether is one of the antirealist's (Laudan in particular) favorite examples, it is worth detailing its observability status.

Everyone, advocate and enemy of the ether alike, agrees that the ether is unobservable. But in what way it is unobservable? The answer to this question comes easily by evaluating the dimensions of ether observability. Probably the most pressing and initially the most interesting dimension is that of immediacy. When Laudan describes the ether as "an entitiy that is regarded as in principle unoberservable" (1981b, 159), does he mean by "in principle unobservable" the same thing that I do, and if so, is he correct in using it to evaluate the observability of the ether? Is the ether unobservable like quark color, or something causally isolated in spacetime, unobservable because it cannot interact in a meaningful, informative way with any observing apparatus? Or is the ether unperceivable in fact, like a photon or an atom? The resolution of this ambiguity begins by consulting the ether theories themselves to see if interaction between the ether and some potential observing apparatus is allowed, and to see, given that such interaction is possible, if information of the ether is conveyed. One looks for information of the ether which comes from the ether, where the source of information is described by the interaction laws of ether theory, and information content is accounted for by that and auxiliary theories.

Not surprisingly, several questions in the analysis of ether observability are similar to those encountered in describing the observability of a photon. Their respective theories describe them both as media of information. The question at hand is whether (and if so, how) they can also be a source of information. With the ether, as with the photon, there is no surer way to find out than to ask the theory itself. I do that through examples of getting ether-information through ether-interaction.

In the object-property ordered pair schema we must look at examples where the object is the ether and the property is some characteristic of the ether such as the size of the particles, their separation, the inter-particle force, the particle motion, or the rigidity of the ether. As one example of such an ordered pair consider the luminiferous ether as it was theorized by Fresnel and Fitzgerald, and the property of rigidity. The luminiferous ether functioned as the medium of light waves in a way analogous to air as the medium of sound waves. Motions in the light source (normal, ponderable material) create, according to the ether theory account, a disturbance in the ether. In this sense then the normal, ponderable stuff of the world in coupled to the ether. Matter can affect the ether. The distrubance in the ether propagates as a wave through the ether, and the ethereal oscillations can excite the eye to cause a sensation of light. So the interaction is effective both from matter to the ether and from the ether to matter. The ether must participate in some interaction with the material of the eye if it is to cause a sensation of light. And in general, if the either is a medium of information it must be able to interact with matter.

The oscillating ether can interact with other kinds of matter besides the eye, as in the demonstration of polarization of light. Here the light waves are directed through one polarizing glass, then through another, identical polarizing glass, the analyzer, and onto a screen. There is interaction between ether and apparatus at least insofar as the quivering ether makes the screen shine with light. As the analyzer is rotated about the optic axis, the screen displays a cycle of brightening to a maximum, dimming to total darkenss, and brightening to the maximum, and so on, completing a full cycle for every 180 degrees of rotation of the analyzer. This behavior of the apparatus indicates that the light is polarized by the first polarizing glass, or, more generally, that light is polariable.

In this apparatus-state information one can find nested ether-information in the following way. From basic apparatus information of analyzer position, brightness of screen, etc., one learns that light is polarizable, information which in turn contains the information that the wave in the ether which causes the phenomenon of light is a transverse wave. Within this wave-information is nested the information that the medium (the ether) must be a solid, because only a solid has the resistance to shearing strain which is necessary for propagation of transverse waves. Thus in the context of the contemporary physical theories, the polarization apparatus recreates information *of* the ether and an ethereal property, namely that the ether is solid. And furthermore, this information derives *from* the ether since it is the product of interaction between the ether and the appartus. The ether acts in this case as both medium and source of information. Experiments such as this which report on the properties of the ether were part of the short-lived science of ether dynamics. The analysis of this example of ether observability, the evaluation of the dimensions of observability, will be postponed only long enough to present another example.

Consider the observability of the caloric ether, the imponderable fluid the presence of which produces heat. An example in this context is immediately different from the previous example of the simple observability of heat Q since now the object-property pair must include caloric as the object, whereas in the other example heat was a property of some other object. This, after all, marks a fundamental difference between caloric theory of heat-as-a-substance and modern kinetic theory of heat-as-a-process. So now we must consider the observability of claoric fluid as the object, with respect to some property such as density, motion, or particle size. A simple example of this focuses on the description of caloric as an ethereal fluid of mutually repelling small particles.

Use an an observing appartus some solid object such as a block of ice. The particles of caloric which, according to the caloric theory are present in the object, will be distributed throughout the object as a result of the replusive force between particles. The caloric particles will surround the particles of the object. (This piece of caloric lore is adapted from a description in Cantor and Hodge 1981, 27-28.) As more caloric is added to the object the ether-cloaked object-particles experience a repulsive force between themselves as the ether-particles push for distance between each other. The more caloric, the greater replusive force and the object responds by expanding. If more caloric is added to the object, the repulsive force of the caloric particles will at some point overcome the binding force of the object-particles and the solid will be seen to change to a liquid. It will be a liquid as long as the expansive force of the ether is great enought to overwhelm the binding force but not so great as to overcome the containing force of atmospheric pressure. Adding more caloric increases the expansive force in the particles of the liquid until, at the point where the expansive force matches the containing pressure of the atmosphere, the liquid changes to gas. The water which was once ice now boils.

The relevant apparatus state then is the expansion and changes of state of the object in response to heating, and using the contemporary physical theories one can find information of the ether nested in the information presented by the apparatus. The information that the object has changed state from solid to liquid contains the information that there is more caloric therein than there was previously. The accounting of this nomic nesting is provided by the caloric theory itself and its description of caloric as a fluid of mutually repelling particles. Hence, one gets information for the caloric, namely that it has flowed into the object. And moreover, this is information which comes from the caloric in the sense that it is an interaction between the observing appartus, the block of ice, and the caloric which is the cause of the apparatus state, its melting, which exhibits the information.

With these two examples in mind we can evaluate the specific dimensions of observability of the ether. The classification of immediacy of ether is apparent, at least that the ether is *not* unobservable in principle as that classification is used here. The relevant theories do not themselves disallow the ether from interacting with matter in such a way as to convey information of the ether. It is therefore unperceivable in fact, but not unobservable in principle. This conclusion leaves unstudied and the possibility that the ether could so interact with some human sense organ and therefore be perceivable but I think this is a safe short cut to make.

That the ether is not unobservable in principle should be easily noticeable on sociological consideration of the ether-scientists' activities. If the theory of ether explicitly precluded meaningful interaction between ether and observing apparatus, then those scientists who understood and believed the theory would not have wasted their time looking for ether or signs of ethereal properties. Michelson and Morley, for example, spent much effort and worked with great credibility in looking for the ether wind. They were looking for specific information of the ether, namely that the ether moves with respect to the earth, and this information would, they theorized, be contained in the information of the state of their apparatus, a state which resulted from interaction with the ether. From the now famous null result of their experiment one can conclude that the ether wind was unobserved, but not that it is unobservable. Were it truly unobservable, reasonable men would likely not have taken the trouble to look for it, and certainly the scientific community would not have paid so much attention, let alone expressed surprise at the null result.

To finish the story of the Michelson-Morley experiment, it is interesting to note that after the experiment left the ether wind in the status of being unobserved, Fitzgerald suggested a revision to the ether theory in which relative motion of the ether causes the observing apparatus to contract in such a way as to just nullify the detection of the ether wind. In other words, the Fitzgerald contraction theory accounts for the unobservedness in terms of unobservability. With the Fitzgerald theory the object-property pair of <ether, relative motion> becomes unobservable in principle, thus explaining its unobservedness. But this is one particular property in only one of many ether theories. It does not indicate that the ether in general is unobservable in principle in this theory. Nor, in all fairness, should this ad hoc patching up be taken as the standard of ether theories.

Evaluating the two examples of ether observation shows that such events can be quite direct in terms of the number of informational interactions required between object and observer. The ether itself interacts with the observing apparatus, the polarizer-analyzer-screen setup, or the block of ice, which then presents the information in a form which is accessible to the observer. Contrast this with the more indirect process of observing the arrival of a photon through its triggering of a multi-stage cascade of electrons whose information must then be processed electronically.

The ether observability also requires a relatively little amount of interpretation to find the ether information nested in the apparatus information. From the news of the polarizability of light waves one concludes, via only one intermediate inference that the waves are transverse, that the ether is solid. Contrast this quick accounting with the more involved unnesting required to find tectonic plate information in the apparatus arrival-time information. There the inference must encounter Q-factors, wave speeds, material temperature, and compositional features of the earth to reconstruct the structural information on the shape of a plate. Compared to this, the amount of interpretation required in ether-observation is minimal.

But, while the amount of interpretation may be reassuringly low, the independence of that interpretation from the ether theory itself is discouragingly low. The informational accounting is thoroughly infected with reliance on the ether theory itself. The theory of the ether is used to account both for the source of information and for the medium of its conveyance in the observation. It is caloric theory which describes the response of the block of ice, for example, to the presence of caloric, and it is the same theory which assures the observer that holding the ice over a flame amounts to adding caloric. The ether theories single-handedly account for the phenomena in which the observer, not surprisingly, finds ether information. At most, only very limited use is made of independent theories, as, for example, when one invokes the conventional description of waves to infer transverse waves from their being polarizable. This is a case, then, of using the very aspects of the theory we would hope to confirm, in cooperation with a meager number of other supporting theories, to account for the reliability of the acquired information. This falls nearly to the bottom of the scale of independece of an account.

It is this lack of independence between theory of observation and theory of object that marks ether observation reports as dubious. By accounting for its own observability and by isolating itself from other relevant theories, the ether theory is prone to the unchecked speculation as to ethereal characteristics. This is Joseph Priestly's complaint:

> Indeed, no other part of the whole compass of philosophy affords so fine a scene for ingenious speculation. Here the imagination may have full play, in conceiving of the manner in which an invisible agent produces an almost infinite variety of visible effects. As the agent is invisible, every philosopher is at liberty to make it whatever he pleases, and ascribe to it such properties and powers as are most convenient for his purpose. (quoted in Laudan 1981b, 159).

But the license for the ingenious speculation comes not simply from the ether being invisible, that is, not preceivable, but from the fact that the only way to get ether-information is essentially reliant on ether-theory.

It is interesting to compare the observability of the luminiferous ether with that of air, since both function as the medium of some perceptual information. One does not hear the air that is vibrating and carrying the sound wave, just as one does not see the ether which is vibrating and carrying the light wave. But with air there are alternative ways to observe which do not require inference to its role as a medium of information. One can, for example, feel the pressure and turbulence of the wind and thereby, relying on the tactile sensation, obtain information on the air which does not depend on the physical laws of air as a medium of sound. This sort of alternative observation is not an option with the ether. The only channel of ether-information is through ether as medium and is essentially supported by the theory of the ether as medium. This lack of independence in the informational accounting is the telling dimension of observability of the ether. You may get some information of the ether but it is entirely self-serving.

There is a kind of incompleteness which plagues the cases of ether observability, much like the incompleteness found in the case of perceiving heat. For a single model of the ether there are usually only very few properties for which the ether-property pair can offer information. This is the case because ether theories tend to be problem-specific in the sense that the ascribed properties of the ether are designed around specific problems, specific phenomena to be explained. The Fresnel theory of luminiferous ether gives the ether enough properties to account for the basic phenomena of light such as polarization and refraction, but not much else. And since it is the ether theory itself which accounts for information, it is only these few properties for which the ether can be said to be only unobservable in fact. It is only a few properties about which one can get information through interaction. In this sense, the ether is observationally very incomplete. This poverty of observation-amenable properties is an important weakness in the observatility in that it contributes to the "ingenious speculation" of those properties which are not amenable to observation, and specialization of ether theories to specific phenomena leads to inconsistency in other domains. The property of solidity, for example, accords well with the polarization of light. It makes little sense though with the fact that the planets and other huge objects must pass uninhibited through the ether.

With the dimensions of observability evaluated, I can summarize the observability of the ether. For one thing, the ether is not unobservable in principle. In terms of the relevant theory on interaction and information exchange, the ether is unobservable in the same way as a photon, that is, it is unperceivable in fact. The telling distinction of the observability of the ether is the low value of both the independence of interpretation and completeness dimensions. These turn out to be the epistemologically important dimensions, the considerations which mark the ether as a dubious entity. In these dimensions is rooted the epistemological disrepute of the ethers to go along with its ontological bankruptcy. The lack of indpendent accounting of the information is a serious reduction of the observability of the ether. So too is the lack of completeness of observability, since that results in an inability to check the large-scale consistency of the theory.

The case of the observability of the ether provides a lesson for studying the observability of other entities, and that is that one must look beyond the image produced by some observing apparatus to evaluate observability. It is necessary to understand how the apparatus-information is produced and how it is related to the object in order to classify the nature of observability. This was suggested earlier in the difference between observing a strand of DNA and observing a single atom using an electron microscope in both cases. The same machine produces similar final images but with importantly different observability classifications. The image of the single atom, in fact, with its lack of independent interpretation, is not far different from an observation of the ether. Conceivably, one could construct a viewing device which imaged the oscillating ether on a television screen in a way similar to the electron microscopic production of a picture of an atom. Set the ether to quivering with a microwave transmitter and let the ether interact with the appropriate electronic circuitry which is connected to an oscilloscope, and the screen will image the oscillations of the ether. The observability of the ether as reported by this ether vibrascope is in the spirit of consulting science, the theory T of the entity, plus other, auxiliary theories, to describe the observability characteristics of the entity. The shortcoming with the vibrascope, as to some extent with electron microscopy of a single atom, is that it relies essentially on T. There is no information about the ether without first using T to plant the information in the account. And the moral of the story is to not rest observability evaluations with what appears on the screen_ check for independent interpretation.

The case of the ether also has a message for manipulability. One can, contrary to Hacking's indication (1983, 275), manipulate ether. This is especially easy with caloric fluid which can be made to flow in or out of systems by changing the temperature, as, for instance, by applying a flame. The rate of flow of caloric is dependent on the gradient of temperature as described by the dispersion equation, $Q/\Delta t = kA\ (\Delta T/\Delta x)$, where $Q/\Delta t$ is the rate of flow of caloric. So by controlling the temperature gradient $\Delta T/\Delta x$, one can manipulate the flow of caloric. Nothing but technological considerations and probable denial of government funding prevents the building of a caloric accelerator to get the flow up to high speeds. Build it two miles long at Stanford, if you like. For the caloric theorist this is in fact flowing caloric, ether manipulated.

I do not intend this fanciful description to be an endorsement of the caloric theory or an encouragement to build caloric accelerators. I do intend for it to point out that manipulability is theory dependent. To find the limits of what is manipulable in the world described by theory T, to adapt van Fraassen's advice, we must inquire into T itself, etc.. And with manipulability, as with observability, an important check on the classification will be in terms of the independence between description of manipulability of the object and the theory of the object itself. The point is that questions of existence are no more easily answered in terms of manipulability than observability. The evaluations of the two are equally complicated and equally susceptable to nepotism in their accounting. If manipulability is to serve as a hallmark of epistemic warrant (as Hacking indicates) than a detailed method of evaluating manipulability is called for.

CHAPTER 4

THE DECLARATION OF INDEPENDENCE

Having presented such a variety of scientific classifications of observability, it will be illustrative to summarize the important features, emphasize the epistemic significance, and distill the moral of the stories.

1. INDIVIDUALITY AND THE LACK OF CLEAR CASES

There is something to be learned immediately from seeing the dissimilarities of the examples and the differences in the meaning of "observable" or "unobservable" in each case. The display of diversity and complexity is valuable as a counsel of caution in forming conclusions on the basis of observability. If we learn nothing else from the many examples we should learn at least to avoid cavalier generalization across observables or unobservables, as in claims to the effect that we have no warrant to believe in an independent existence of unobservables. The examples demonstrate that the bare term "unobservable" hides important differences and indiosyncracies. A glance back at the examples verifies this. Considering the quantum mechanical entities described, there is no univocal sense to an expression like "the unobservables of quantum mechanics". There are differences in the details of observability between quantum states and quantum particles such as quarks or photons, and what it would mean to observe each is importantly distinctive. And within the category of quantum particles, observability characteristics are dramatically dissimilar. A single quark is constrained by color confinement into a different, stricter kind of unobservability than a photon. And even within the concept of a quark, different properties must be described by different observability characteristics. Quark color, it has been shown, is unobservable in a way that quark flavor is not.

It has been shown too that even within a single technique of observing, electron microscopy, there are relevant differences in the evaluation of observability samples. As imaged by an electron microscope, the observablity characterisitcs of a strand of DNA differ from those of a single atom in terms of the nomic nepotism required to unnest the information.

The point is that the cases surveyed are importantly individual in their observability characteristics. The term "unobservable" is virtually meaningless without at least a sketch of an analysis of the particular case. "Unobservable in what way?" is the appropriate and helpful response. If a clear case of observability is one whose characteristics are apparent with little or no analysis, that is, if clarity of cases is measured by complexity of analysis, then there are very few clear cases, and maybe none at all. Surely none of the examples surveyed here are clear cases of observability. And these examples have been purposefully chosen to represent important entities in scientific theories. Quarks, state-functions, photons, tectonic plates, heat, acceleration, and bubble chamber results are all entities which function essentially in scientific theories and practice, and the nature of the observability of each is distinctive and complex. None of these important cases is a clear case of observablity, and it is a relatively safe inductive step to say that all, or at least most, important cases will be complex and distinctive. And because of this one cannot rest a description of the nature of observability with an attitude which acknowledges the possibility of vague and complex cases but ignores their importance.

The diversity of examples also points out that there are good and bad ways, that is, productive and unproductive ways, to put questions of observablity. Of an entity x, a photon or a softball, say, questions of the form "Is x observable?" or "Is x unobservable?" are questions badly put. Such questions are asking for trouble in that the short answer they seek will hide the relevant complexities and diversity of observability. The short answer is inadequate to service the issue of realism. The examples and the interaction-information schema point out a more informative way to put the question of observability. For one thing, it is better, more productive for learning about the nature of information from experience, to evaluate observability of object-property pairs. Better to ask of the observablity of object x with respect to property P. And it better suits the subject to ask in a way that does not seek the short, prescriptive answer. Ask instead a question of the form "In what respect is $<x,P>$ observable or unobservable?" Even better, put the question as, "What are the observability characterisatics of $<x,P>$?" This approach allows the full, complex nature of the observability of $<x,P>$ to emerge.

This approach which demands the longer answer is better suited to respect the important points as raised by both the realist

(Grover Maxwell) and the anti-realist (Bas van Fraassen). The realist convincingly points out that there is no clear dichotomy between observables and unobservables as the bad question and its short answer must assume. And the anti-realist responds, equally convincingly, that even without the sharp dichotomy there are some important differences in the observability of scientific entities, differences which should be brought out and made to speak on the issue of scientific realism. Features of observability ought to indicate something about the justification for our belief in the existence of things. Observability surely has some epistemic significance and some relevance to the issue of scientific realism, even if there is no clear dichotomy. The long answer to the observability question respects both of these attitudes. Without forcing a dichotomy it seeks the epistemic significance of observability.

There is another way to put the distinction between productive and unproductive approaches to observability, a way which highlights the empirical basis of the issue. The strategy is to let science decide or at least significantly contribute to the decision of observability. But this does not require, and in fact one should avoid, looking to science for a dichotomy in terms of what *qualifies* as observable and what does not. This would be the theme behind a formulation of the question of observability which consults the relevant science to determine whether it is only unaided perception which *counts* as observation, or whether images produced with microscopes or radio telescopes can count as observable as well. Such an approach pushes us toward a dichotomy in observability into which one is prone to read unwarranted importance.

It is better to ask of science what information of the object-property pair <x,P> comes to the scientist from x. How does the information come and how is it recognized as information about x and P? It is advisable even to phrase observability questions (and answers) such as to avoid the terms "observable" and "unobservable". The more productive attitude is one which looks to see what goes on in cases of observability rather than to see what qualifies in some a priori classification. Then the results, the long, descriptive answer, can be applied to the issue of realism. For a particular case <x,P>, ask what it all, the whole story of <x,P>-observability, means about the warrant for belief in the existence of x. And more generally, ask what the observabilty characteristics in general, the dimensions of observability, indicate about the justification of a realist interpretation of scientific theories.

The importance of the interaction-information picture of observability and the description of a multidimensional observability space is not to extend the category of observable beyond the directly sensible and into the microrealm of electron microscopy and elementry particles. Nor is it to lobby for a broader application of the label "observable". The intent of the method is rather to resist the urge of such categorizing and labelling as a way to understand observability and its role in the issue of realism. The procedure is to disunify the classification of observability by demonstrating, through an empirical look at what the sciences say, the many degrees of freedom in the concept of observability, and the important differences in the variety of examples. The disunity and diversity though, do not mean that there are no implications which can be drawn from a study of observability for the issue of realism. We have already noted the implied caution against careless generalization across unobservables. But there is more to be said than that. Pointing out that the concept of observability is multidimensional and diverse does not render it meaningless. To assess the impact of a study of observability on the issue of realism we should go beyond the claim that observability is multidimensional and describe what each of the dimensions means. That is, what do values along each of the dimensions indicate about the epistemic status of <x,P>? What is the epistemic significance of each dimension? The description of observability of any pair <x,P> is typically long and complicated, and to find the imlications for the issue of realism one must ask what it *all* means.

2. WHAT THE DIMENSIONS MEAN

To this end we can reassess the description of the dimensions of observability as illuminated and refined by the details of the examples. Consider *immediacy* which reports on the possibility of informative interaction between the entity and some apparatus for observation. The three categories of immediacy roughly reflect intervals in the continuum of directness. The distinction between things unperceivable in fact and things simply perceivable is also sensitive to the fact of whether the observing apparatus is human or not. The category of principled unobservability, those things which simply cannot interact and convey information, no matter how indirect the interaction or how deeply nested the information, represents the clearest distinction in the survey. If there truly are such things, that is, if my analyses of quarks and things causually isolated in spacetime are accurate according to relevant theories (and I think they are), then any interpretation of science must reckon with things which are informationally isolated.

But the immediacy evaluation is just a heuristic *triage*, and the examples demonstrate that it is too coarse-grained to be epistemically significant. The reliability of information-acquisition reports varies importantly within each of the immediacy categories and there are even some cases of perceivable object-property pairs which we are inclined to regard as less reliably observable than some cases in the unperceivable-in-fact category. The immediacy dimension simply does not reveal what is epistemically significant in observabilty. That is, of course, old news, and only repeats the claim of those who argue against a clear dichotomy between observables and unobservables. But it is a claim worth emphasizing with the examples of chapter three.

A more fine-grained approach to evaluating the physical dimension of observability is the measure of *directness*. This is an indication of the causal distance between object of perception and human observer since it measures the amount of interaction, or more exactly the number of interactions, between the object x and the state in which x-information is rendered in a form which is accessible to the human observer. Directness of observability measures the length and complexity of the causal chain from x to the observer. It reports not only the number of intermediate entities which carry the information from the source to observer, the photons, seismic waves, electrons, or whatever, but also the interactions they encounter. Electrons whose paths are bent in magnetic fields and interfered with by other electrons carry information in a less direct way than electrons uninhibited. But directness ignores any arbitrary decision as to what is to count as an observing apparatus. It is unconcerned, for example, whether in doing seismic tomography one regards only the seismometers as observing apparatus, or extends the apparatus to include the entire interior of the earth as a seismic lens. What matters to directness is the total length of the interaction chain, and not the length to whatever is considered to be the apparatus.

The case studies demonstrate that the possible values of the measure of directness can vary dramatically. Observability can be as direct as the process of thermodynamic heat in which there is the single interaction between the object and the contiguous human sensory receptor. With heat there are no intermediate interactions necessary to convey the information from object to observer. More typical though is the case of observability of acceleration of a macroscopic object. In this example, as with most naked-eye visibility examples, there is interaction between objects and photons, and then between photons and body. Here the object is causally distant by one intermadiate of information, the field of photons. An example of a longer chain, that is, of less directness in observability, is in the conventional transmission electron microscope. In this case, information originating in the specimen is transported through interaction between specimen and electrons, and then electrons and photon-sensitive screen, and thence via photons to the human body. A scanning transmission electron microscope is even more indirect. Again the information takes its first step from the specimen with an interaction between specimen and electrons. The path is long though, requiring interaction between the electrons and a detector which produces an electrostatic potential, instigating interactions with electrons in the wires of circuitry, producing a current... . The chain continues to the electrons projected onto the scintillation screen of a television, to photons and the eye. Even without an exact method for counting interactions, it is clear that the physical path of information is interactively longer in the case of the scanning electron microscope than it is in that of the conventional electron microscope. In this sense, the specimen, the object of observation (or, less tendentiously, the source of information) is causally more distant in the former case. Yet even more indirect is the observability of a photon using a photo-multiplier tube. The photon initiates the interaction sequence which carries information of the photon by interacting with a single electron at the detecting end of the tube. This primary electron then interacts with other, secondary electrons in what is the first in a lengthy sequence of electron interactions, The informaiton is carried then through interaction with a detector, electronic circuitry, and eventually to the observer's eye.

The measure of directness in observability can often be disguised by the packaging of the instrumentation. The photomultiplier looks compact and direct as a processor of photon information, whereas the collecting of plate tectonic information using seismic tomography looks to be a huge, vastly indirect mess. But as has been pointed out before, this appearance is a result of the size of the specimen under observation and is not necessariliy related to the length of the interaction chain involved. If anything, the observability of the tectonic plate is the more direct.

While directness is an indicator of a sense in which the object is physically distant from the observer, the *amount of interpertation* answers for its epistemic distance, This is measured in terms of how deeply nested the information of the object is in the final state information available to the observer, that is, how much inference must be invoked to get from the observer-state information to the object-state information. This is not necessarily correlated to the naturalness, ease, or speed of the inference. This is not even to imply that the inference is in fact done by the observer. The inference we speak of here is that justification which is available to the epistemic community should an explicit account of the informational content be called for. For example, in the case of heat, the inference as detailed in chapter three is not an immediate or easy step for the observer to make, but it is a valid and theoretically uncomplicated inference to make according to the relevant science being used to report on observavility. The inference is not immediate or natural perhaps because "heat", as it is used in the example, is a theoretical term whose meaning and relation to other terms is governed by a somewhat unnatural theory. In the case of acceleration, on the other hand, a concept which is less dependent on theory for its specification, the inference from observer-state information to object-state information is easy and fast. But even in this case where the correlation is natural (perhaps naturally selected), the information must be inferentially unnested.

Like directness, amount of interpretation is seen to vary all over the place. Very little inference is involved in finding object-acceleration information nested in the state of motion-recpetors in the visual system. More interpretation is needed to find information of a photon in the report of a photomultiplier. One must account for the initial photon-ionization, the release of secondary electrons, and the resulting pulse of current on the electronics. And perhaps even more inference is called for in unnesting plate tectonic information in seismic tomography. From the characteristics of wave interaction one infers local wave speeds and thence to local temperatures and composition, to age of material, and to structural features.

Amount of interpretation can also be disguised by the packaging of the tools. An electron microscope produces, sometimes, an easily recognizable image. Features of the image can be readily noticed because the full processes of interpretation are built into the machine. This contrasts with the case of seismic tomography in which the procedures of interpretation are open and obvious and in which the images produced are not easily recognizable for what they are supposed to be. The features are not readily noticeable. But this is no indication of the amount of complexity of informational interpretation in either case. It is indicative only of the possibility of building the interpretive process into a neat machine and concealing it from the casual viewer.

The dimension of amount of interpretation reports on the quantity of inference but not on the quality. An important aspect of this oversight is filled in by evaluating the *independence of interpretation* which reflects the closeness of relation between the theory of the object in question and the theories used to account for the information. The possible values of independence vary significantly form one case to another. In the case of using electron microscopy to convey information of a plant cell or DNA, the physics which accounts for the nomic nesting of information is quite independent of the biology of the specimen. Less independent is the case of electron microscopy of an individual atom, since in this case the physics of scatering and electron optics are related to atomic physics. The epistemic significance of the relation is that the theory of the atom is being given an influential vote in the decision of observability of its own entity. But the least independently accountable case of observability presented here is that of the ether. The theory of the ether is given virtually the only vote in adjudicating the observability of the ether and so there is no independence at all between the theory of the object and the laws which account for observability.

The details of independence of interpretation as outlined in chapter two (and summarized on Table 1) facilitate a detailed account of relative scaling of the examples along this dimension. The strongest case of independence is such that no part of the theory of the object is used in the interaction-information account. Of the examples presented, the observability of acceleration is the closest to this ideal. If T_x is classical mechanics, there is no part of T_x that appears in the observability account. At the other extreme is the case of the ether in which that part of the theory which describes the properties about which we are allegedly getting information from the world is nearly the entire basis of justification of the alleged information. In these two extremes, the measure if independence clearly parallels our intuitions about observability, an intuition which accepts acceleration information as a reliable report from the world, but rejects ether information as specious.

Between the two extremes the details of independence instruct us to identify parts of the theory of the object, the part of the theory of DNA (or plate tectonics) which describes shape and the part which describes composition. Following this plan we can scale the observability of the shape of DNA molecule, as imaged by an electron microscope, as being the more independent account than the observability of a single atom, as imaged by the same machine. Any example of an observability report can be fit into the scale in this way and the measure of independence can be useful in educating intuition regarding the reliability of these informational reports.

There is one candidate for a relevant dimension of observability which I do not consider to be important. That is noticeability, and it is probably already apparent from earlier disparaging suggestions why I think it lacks the epistemic significance to be included as a dimension. The ease with which a feature is noticed, either by the trained or untrained observer, is too dramatically a product of the packaging of the instrumentation. Recall the example of DNA where features were displayed either in an easily noticeable form on the television screen of an electron microscope, or in a less accessible way as the result of biochemical analysis. The difference in noticeability could easily be eliminated with the appropriate transducers and a television screen in the biochemist's lab. The information that the strand is a closed loop (if that is the property in question) resides in both results, and processing the information into a noticeable form will not add more information of the object. For this reason, noticeability, while a convenient feature, is not epistemically significant.

In using the survey of examples to display the variability of the values of the dimensions of observability, it is apparent that, except for the relation between immediacy and directness, the values along the different dimensions vary independently. In other words, the dimensions of observability are orthogonal. There is no correlation, for example, between the measure of directness of observability and the amount of interpretation. Physical distance is not necessarily related to epistemic distance. In the case of photomultiplier observability of a photon the information is passed through many interactions as the electrons cascade through the tube, but since all of these interactions are the same, there is only one inferential step required to account for the informational exchange. What is vastly indirect can also be epistemically brief and uncomplicated. So too can what is largely indirectly observable involve a large amount of interpretation, as in the case of the observability of tectonic plates. And, of course, what is physically very direct is not always epistemically straightforward, as in the case of heat. Even though the interaction is directly between object and observer, the imparted information must be inferentially unpacked. The point is that for any particular case, physical directness is no indication of epistemic accessibility of information, nor vice versa. These dimensions are orthogonal.

It is the same for the amount of interpretation and its idependence; the value of one is not indicative of the value of the other. This orthogonality is plausible a priori since there is no reason that the use of a large number of theories to do the informational accounting should imply either that one of those theories is or is not the theory of the object. Furthermore, a very brief, one step inference could be utterly without independence if that one step relies on the theory of the object. The evidence bears out this orthogonality. The observability of a tectonic plate involves a great deal of inference to find plate information in the final velocity-anomaly information, yet the theoretical support for the inference is largely independent of the plate theory itself. The case is reversed with the ether, where observability includes only minimal inference but that which is required draws solely on the theory of the ether itself. And there are the two examples from electron microscopy, the observability of an atom and of DNA. They require roughly similar amounts of interpretation to reveal information about the specimen, but the degrees of independence of inference are importantly different between the cases.

Not surpisingly, there is no correspondence between independence and directness either. Observability of tectonic plates, which is highly independent, is quite indirect. Observability of a single atom is also quite indirect but significantly less independent.

The orthogonality of the dimensions is an important feature of observability space. It indicates that the values of the dimensions vary independently and hence that the epistemic significance of each dimension must be assessed individually. Something is epistemically significant to the extent that it is a good indicator of reliability of belief, in this case, belief in an information-acquisition report. Our situation as philosophers and scientists makes available to us the informational claims about the world together with the interaction-information evaluation of the dimensions of those claims. And it is from this perspective internal to science and reason, that is, without direct access to facts about the make-up of the world, that we must decide what the world is like. An important question then is, of these dimensions of observability, which, if any, should instruct us as to which reports about the world to believe. The orthogonality of the dimensions indicates that we ought not talk about the epistemic significance of observability. We should ask instead about the epistemic significance of each dimension.

The degree of directness indicates the physical length, as measured by currently accepted scientific theories, of the interaction chain of information. The examples of the observability of the ether and of heat both demonstrate that a good score on directness (few interactions) is not a sufficient indication of a reliable information report. Something is suspicious in the claim of ether observability, but it is not a lack of directness.

In the context of the interaction-information account, a good score on directness is not a necessary indication of reliability either. What counts in the interaction-information account is that the information from the object arrives in the final apparatus- or observer-state. In all cases, the laws we use to describe the flow of information indicate that the information does in fact pass through each interaction. That is, we are not dealing with cases of laws describing probability of informational exchange, cases in which more exchanges (more interactions) increase the chances of lost information. Only in such a case would the length of the chain of interaction be an indication of the reliability of the acquisition of information. But where the laws specify that the information is in fact exchanged (citing conservation of momentum or energy, for example) then a high number of exchanges (a bad score on directness) does not indicate unreliability. Since our only available assessment of the interaction-information process and its directness is through laws of nature, it is in the use of laws that we must find epistemic significance.

This dimension of directness has often been cast as the central issue in observability. Grover Maxwell (1962) takes us through several degrees of directness, from observations through air, plane glass, lenses and so on. Van Fraassen (1981) claims that something is observable (in our terms, the information is reliable) only if the majority of any indirectness it might involve can be avoided under some circumstances. But Maxwell's account is persuasive because we know that information of the world does pass reliably through air, plate glass and even lenses. And a few more lenses (a little less directness) does not threaten this reliability. We believe this because the relevant scientific description claims that the information *does* pass, not that it *might* pass. Since, according to the relevant, available description, each interaction does reliably pass the information, the number of interactions cannot be significant for assessing epistemic reliability.

What can be said of the epistemic significance of the dimension of amount of interpretation? Here again, whatever claims are to be made should be motivated by and consistent with the examples of chapter three. As with directness, this evidence indicates that the amount of interpretation is not a great source of epistemic significance. And again it is the observability of the ether which suggests that a small amount of interpretation is not sufficient as a mark of reliability of an informational account. There is relatively little interpretation invoked in unnesting ether information. Hence, insofar as we regard the observability of the ether as epistemically deficient, a small amount of interpretation does not indicate a reliable account.

Other examples such as of the objects imaged by an electron microscope show that a relatively lower amount of interpretation is not necessary for a relatively more reliable account either. Beliefs about the shape of a DNA molecule as acquired using an electron microscope seem to be relatively more justified than similarly acquired beliefs about single atoms, even though the amount of interpretation is roughly the same in each case. And certainly we regard the information of the DNA molecule as more reliable than the information of the ether as described in chapter three, even though the DNA case involves a larger amount of interpretation than the ether case. Thus a small amount of interpretation is not a necessary indication of epistemic reliability.

This too makes sense when you consider our perspective for evaluating epistemic significance. We must believe that any of the theories used in an account *could* be falsely describing an interaction-information event. But it is a somewhat flat-footed way of assessing reliability to claim that the use of more theoretical support introduces more risk of error, and less theorizing brings less risk. Granted this may provide some measure of unreliability in the sense that if we know *only* that *some* link in the chain may fail, then it is best to use fewer links. But this assumes that we have no means of identifying the suspicious link, that is, no means of evaluating the quality of an inferential link. Evaluating the quantity of inference, the epistemic length of the chain, is then valuable as a fall-back only in case we cannot evaluate the quality of the inference. But in the interaction-information account there is a measurement of quality. We can use the independence of interpretation to catch the suspicious links in the chain. In this context then, the amount of interpretation has relatively less epistemic significance than does the independence of interpretation.

What counts (epistemologically) in the interaction-information account is that the information of the object-property pair comes from the object and arrives at the observer in such a form that the object information can be unnested. How many interactions are required to transport the information and how much inference is involved in its unpacking are relatively less important than independence of interpretation. An intuitive assessment of the examples, particularly the cases of the electron microscope and the ether, suggests that what is important is independence in the account. Since the information must be unpacked, the reliability of the unpacking must be accountable, and that accountability comes as the independence between the causal laws used in analyzing the information and the theory of the object itself. The epistemic reliability can come only with disinterest such that the theory describing observability has little or no important stake in the outcome.

In this spirit, the laws of visual perception have no dependence on the kinematics of acceleration, and we are happy to acknowledge that the eye delivers acceleration information. Similarly, electron optics and electron scattering theories do not rise or fall with the observability of DNA, and so they provide an independent accounting of the flow of DNA information to the observer. Contrast this with the observability of the ether where we still have causal laws which point to ether information in the effects. But the causal laws are strictly ether laws. The ether theory describes the effects of ether and portrays at least some of them as being able to interact with and convey information to a human observer. But we are loath to call this kind of effect an observation because the entire interaction chain is described and assigned informaitonal content by the theory of the ether. It is this kind of nepotism in the accounting which impeaches the information.

Consider Oliver Lodge's apology for the invisibility of the ether: "It is commonly said that we have no sense organ for the appreciation of the Ether; and we have not any means of appreciating it directly, but we are very much accustomed to appreciate the phenomena which go on in it, in other words to apprehend its modifications" (1925, 26). His message is that though we cannot observe the ether, we can and do observe its effects. But the shortfall in ether observability is not that we can only get as close as effects, for effects might well hold information of the object. The weakness is that, of the effect we can see, there are no *independent* effects, no effects whose ether information content is accounted for by something other than the theory of the ether.

The point is that the important dimension is the independence of interpretation. It is important as a measure of reliability in the account of information of an object. It bears the burden of epistemic significance of observability as an aspect of scientific realism.

3. INFERENCE TO THE SOURCE OF INFORMATION

So far I have argued that insofar as observabilty is to be an important factor in scientific realism, the dimension of independence of interpretation should be weighted more heavily than the dimensions of directness or amount of interpretation. The implication for scientific realism of the result of the independence-weighted interaction-information account of observability can be stated briefly as an inference to the source of information. This speaks to the specific issue of realism of scientific entities (as opposed to realism of theories), where the question is over the warrant for belief in the independent existence of entities as a function of their observability. When observability is described in terms of the information *from* x and of x, there is no sense to the discussion unless x exists. The reasoning here is closely analogous to Nancy Cartwright's "inference to the most probable cause" (1983, 92). The interaction-information account, like a causal account, has the existential claim built in in the sense that if you accept the fact that there is information about x which comes from x, then you also get the fact of the existence of x, existence of the source of information. This idea requires further explanation, and in doing so I will follow Cartwright's clear account of inference to the most probable cause.

The core of the argument for entity realism as it is opposed by an empiricist such as van Fraassen is the inference to the best explanation. In short, we have warrant for believing in the existence of some unobservable entity x insofar as x figures essentially in the best explanation of observable phenomena. Van Fraassen's (1980) focused response to this abductive inference is to ask what explanation has to do with existence. The existential claim about x, and any claim about the truth of x-statements, are gratuitous to the explanation of phenomena, since the claims about x could function equally well in an explanation whether x's existed or not.

Cartwright, in response to van Fraassen's disassociation of existence from explanation, points out a special subclass of explanations, causal explanations, to argue for entity realism.

Her point is that causal explanations are unique in that they automatically include the claim of existence of the entity. If one says about unobservable x that the properties of x explain some phenomenon, this claim is uncommitted to the existence of x. But saying that x *causes* the phenomenon is including an existential claim about x, because x must exist to be an effective cause. For example, quarks and quark properties may be invoked as the best explanation of the patterns in populations of hadrons, but the quark theory may nonetheless be only a heuristic tool for understanding elementary particles. The explanatory success of quarks does not establish their existence, no matter how certain one is of that explanatory success. But where the explanation is explicitly causal, if, for example, one did a Millikan-type experiment and identified fractional electrical charges at work and explained the phenomena as being caused by quarks, then the explanation includes the claim of existence. The quarks must exist to cause observed events. Insofar as you are certain of the causal account you must be equally certain of the existence of the causal agent.

The interaction-information account of observability implies a result much like Cartwright's. An object must exist if it is to be a source of information. The chain of interaction is a causal chain and in this sense Cartwright's claims apply directly. But inference to the source of information is somewhat stricter than inference to the most probable cause. A source of information must be a cause; it must be that the final state of the observer is caused, at least distantly caused, by the object. I have been emphasizing this by saying that the information must be *from* the object. But a source of information must be more than a cause; it must be a meaningful cause in the sense that the effect contains information *of* the object. Recall the example of the photon as it interacts with the human eye. The photon is the cause of the resulting retinal state but it is not a source of information since, as argued in chapter three, the retinal state does not contain information *of* the photon. In this case then, inference to the most probable cause would license an inference to the existence of the photon, but inference to the source of information would not. The difference is in the strictness of the inference, to cause or to source of information. Either inference, once accepted, carries with it the existential rider.

All of this discussion is set on a stage upon which the acceptance of the account of the cause or source of information is assumed. We have not dealt with the issue of the criteria of acceptance of the account either of the best cause or the source of information. Cartwright is very brief on this question of the epistemic markings of a good causal explanation. Her concern is only in pointing out the existential punch of, "the inferences you can make once you have accepted that [causal] explanation" (1983, 93). She separates the issue of what should count as a good causal explanation from what you can do once you have one, by leaving the first problem in God's lap. If God tells us that (that is, we are certain that) quarks explain characteristics of hadrons, we are no closer to knowing whether or not quarks exist. But if God tells us that photons cause the retinal reaction of vision, then he has also told us that photons exist. And by interaction-information standards, if God tells us that tectonic plates are a source of information, then we learn as well that tectonic plates exist.

This is all good news as far as it goes, but of course God does not go around telling us these things. There is a serious incompleteness in the argument of realism with respect to entities x as causes of observable phenomena until we know *how to tell* that x is in fact a cause. The same obligation applies to the inference to the source of information. We now want to know what sanctions the inference about the cause or the source of information so that we can measure the credibility of the corresponding existential claims and commit out belief accordingly. In the case of the causal explanation we must consult the relevant causal laws, and here it seems that the dichotomy between realism about entities (whether or not they exist) and realism about theories (whether or not they are true in what they say about unobservable entities) is blurred. To evaluate the inference to x as cause, one must evaluate the truth of at least that part of x-theory which descirbes the causal powers of x. In other words, to complete the argument for the existence of the entity x (or for entity-realism in general) we must invoke an argument for the truth of that part of x-theory which portrays x in its role as a cause (in general, the causal accounting depertment of the theory must report truths). This much complicity between theories and entities must be admitted to complete the argument.

The interaction-information account of observability provides a mechanism to fill in the epistemic evaluation of the inference as to what is the source of information. As with a causal explanation, an account of interaction and transmission of information is done in terms of physical laws. The game is to consult the applicable scientific theories to describe observability. The best standard by which to measure such accounts is the dimension of independence of interpretation. An inference that x is a source of information is a good one insofar as the inference ticket is issued by an independent source, a theory with no stake in the specific properties or existence of x. Lacking the word of God, independence of interpretation is the next best measure.

It is important to make clear that there are two inferential steps being discussed in this approach to realism. One is the inference *that* x is an explanation, cause or source of information. The other is the infernece to the existence of explanation, cause or source of information. This second inference is the easier since it is a once-and-for-all argument about explanations, causes or sources of information in general. It trades on general ontological features of what it is to be an explanation, cause or source of information, but does not rely on contingent features of the world. Actual claims about features of the world, that this causes that, are made through the first inference.

Cartwright argues in behalf of the second inference in the special case of inference to the existence of cause, but against the second inference in the more general case of inference to the best explanation. My strategy has been to adapt Cartwright's argument in this second inference and to propose an inference to the source of information. But it is in the first inference where the importance of independence of interpretation arises. To judge the acceptability of the inference *that* x is the source of information, one evaluates the independence between the x-theory and the theoretical account of the observability of x. It is with this first inference that Cartwright is brief and does not indicate how to discriminate good from bad inferences that x is the cause of some particular phenomenon. It is unlikely that she can rely on a measure of independence since the causal activities of x will be described by x-theory alone. Causal explanations, like explanations in general, are less suited to the evaluation by independence of interpretation than is the interaction-information account of observability.

4. THE BIG PICTURE

So far the account of observability, the role of science in accounting for information, and the importance of independence, has been kept localized to one pair <x,P> at a time. We have outline how the observability of a particular object-property pair is supported by a set of theories $\{T_i\}$, and pointed out that what is important is that the T_i be independent and reliable theories. Both independence and reliability are part of a larger picture, namely the entire network of entities and properties used in confirming the theories used in accounting for observability of entities and properties. The observability of <x,P> is accounted for by $\{T_i\}$, and x and P then function in T such that the observability of <x,P> is an aspect of the confirmation of T. Each of the theories T_i used in accounting is also subject to confirmation with respect to some entities x_i and properties P_i whose observabilty is accounted for by other theories, and so on.

This suggests as branching network, spreading from the <x,P> of initial concern through levels of theories, entities and observations.

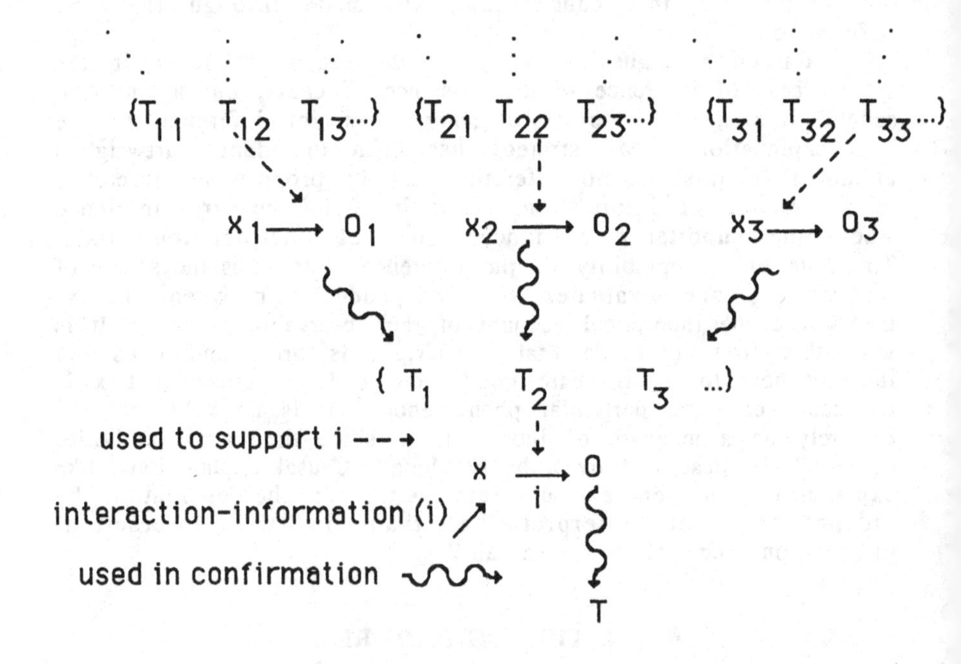

This is not meant to suggest that an evaluation of a humble report that x is P demands an assessment of such a network. It is only intended to point out the place of observation in the world of epistemology of science. Observations not only support confirmations (of the theories found below the observation in the network), but they are also supported by confirmations (of the theories found higher up in the network). To know that the accounting theories $\{T_i\}$ are reliable, as well as independent, one would have to investigate the confirmation of those theories. That is, one would have to analyze the confirmation link ($\rightsquigarrow$) in the network. But this is another issue and it is a project too large to confront here. The primary concern while analyzing observation, considered as the acquisition of information, is to identify which reliable theories are used in the account and to see that they are used properly, that is, independently. This propriety in the account of information acquisition functions as a relative scale of warrant for belief.

The point is that recognizing the generalization that theory influences observation and our acquisition of information is only the beginning of an analysis of observability and its significance for epistemic warrant for belief. One wants to know just *which* theories are influential and to what extent their influence threatens the reliability of the information. The examples of the previous chapter suggest that independence if interpretation (under the hypothesis of reliable theories of accounting) is an accessible measure of this threat. It is then an assessment of the hypothesis of reliability which calls for a description of the full network.

5. SUMMARY OF CHAPTER FOUR

Here is what I hope to have shown in this chapter. The immediate payoff from the lengthy survey of examples of observability is a warning against generalization. The decisive factors in the observability of a photon are sufficiently different from those in the observabilty of quarks, the quantum state-function, tectonic plates or heat to make any claim about unobservables in general ill-advised. And in the effort to find what is epistemically significant about observabilty, the examples were used to show that the dimensions are mutually orthogonal and that the dimensions of directness and amount of interpretation are of little epistemic importance relative to the independence of interpretation.

A further demonstration and description of the importance of independence was provided in the case of inference to the source of information. Not only do I, following Cartwright, argue that knowledge of the source of information gives you knowledge of existence, but also, unrelated to Cartwright's inference to the most probable cause, that independence of inference is an indication of warrant to believe in an account of a source of information. The result is a relocation of emphasis in the question of scientific realism from worry about warrant for belief as a function of observability to worry about warrant for belief as a function of independence, one of the dimensions of observability.

SUMMARY

Here is what I hope to have shown in this book.

The interaction-information account of observability as presented in chapter two and put into action throughout chapter three is intended to be a method of understanding the problem of observability in science, if not a solution to it. It is a method which respects the valuable points and distinctions drawn in previous studies of observabilty, outlined here in chapter one. The realist is correct to insist that observability is more than a simple dichotomy between observables and unobservables. And the empiricist is right to regard observabilty as epistemically significant as a factor in warrant for belief in physical things. The interaction-information account respects both of these claims by seeking the epistemic significance of observability while remaining sensitive to its complexity.

The interaction-information account is how to use science to study observability. The examples of chapter three show how it is done and what can be learned about information from the world when one resists the urge to categorize and concentrates instead on describing the action. Chapter three is a tour of observability space, pointing out the variability of the values of the dimensions and describing some of the important occupants of the different regions, photons, quarks, accelerating objects and so on. The descriptions are informative for their own sake, to know about the availability of information from and of these things, but also informative collectively as a map of observability space.

The tour culminated in the epistemically important aspects of observability. Principle amoung these is independence in the account of interaction. The concept of independence of interpretation is explicated in terms of parts of the theory of the object-property pair in question, and participation in the account of observability. Observability reports represent an intersection of theories other than the subtheory which describes the object itself. And with independence as a measure of this otherness in the observabilty account we also get a measure of epistemic warrant for trusting claims about a source of information. Together with the implicit inference to the source of information, this is the implication of the interaction-information account for the issue of scientific realism.

This study of observability as a preface to the issue of scientific realism has proceeded by easing the notion of observabilty off the stage, it being too suggestive of a dichotomy or, in its more liberal interpretation, of a one dimensional continuum. The focus instead has been on the process of acquiring information from the world. With this approach we come to relocate the emphasis in the realism issue from observability to one of the dimensions of observability, the independence of interpretation.

The interaction-information account, as a method for studying the epistemology of science, is prepared to do the study from a perspective internal to science. Since science is in the business of determining what there is in the world, such determinations must be made in part on the basis of the characteristics of our scientific knowledge of things. And a study of how we know about things cannot help itself to claims about what exists, since those existential claims are supported by the epistemological evaluations, observability evaluations, for example. That is, the epistemology must be done, at least in part, internal to science. But it need not be confined to being internal to a single theory such as the theory of the entity in question. To account for the observability of x, one takes advantage of a perspective external to the theory of x, that is, an account by independent theories.

The dimension of independence bears a heavy burden in the interaction-information account of observability. The eclectic nature of observability evaluations, evidenced by the fact that there are few direct reports of observability in science because observability does not fall in the domain of any one theory, is insurance that the evaluation will not be an artifact of a theory. But could it be an artifact of a conspriacy among theories? Perhaps electron theory and electrodynamics take in the wash for biochemistry in the description of observability of DNA. Perhaps optics renders a similar service for kinematics in describing the observability of acceleration. And maybe even our common sense notions of what is observed does the laundry for all of science. Maybe. But insofar as epistemology of science must be done from as internal perspective, that is, without prior information on the ontology of the physical world, we need to find criteria for epistemological evaluation within science itself. Lacking the word of God or advance knowledge of what there is in the world, independence in the interaction-information account is the surest epistemological foothold. In any case, the wash must be done, and in epistemology as in life, better someone else does it than do it yourself.

REFERENCES

Achinstein, P. (1968): *Concepts of Science*, (Johns Hopkins Press, Baltimore).

Achinstein, P. and Hannaway, O. (eds) (1985): *Observation, Experiment and Hypothesis in Modern Physical Science*, (MIT Press, Cambridge).

Albright, J. (1982): "Comments Concerning the Visual Acuity of Quark Hunters," *Synthese* **50**, 147-152.

Anderson, D. and Dziewonski, A. (1984): "Seismic Tomography," *Scientific American*, October 1984, 60-68.

Ballam, J. and Watt, R. (1977): "Hybrid Bubble Chamber Systems," *Annual Review of Nuclear Science* **27**, 75-138.

Belinfante, F. (1975): *Measurement and Time Reversal in Objective Quantum Mechanics*, (Pergamon, Oxford).

Blokhintsev, D. (1968): *The Philosophy of Quantum Mechanics*, (Reidel, Dordrecht).

Bohm, D. (1951): *Quantum Theory*, (Prentice Hall, Englewood Cliffs).

Bott, M. (1982): *The Interior of the Earth: its structure, constitution and evolution*, (Elsevier Science Publishing Co., Inc., New York).

Boyd, R. (1984): "The Current Status of Scientific Realism," in J. Leplin (ed.), *Scientific Realism*, (University of California Press, Berkeley).

Brindley, G.S. (1970): *Physiology of the Retina and Visual Pathway*, second edition, (Edward Arnold LTD, London).

Cantor, G. and Hodge, H. (eds.) (1981): *Conceptions of Ether*, (Cambridge University Press, Cambridge).

Carnap, R. (1936): "Testability and Meaning," *Philosophy of Science* **3**, 420-68, and **4**, 1-40.

________ (1956): "Methodological Character of Theoretical Concepts," in Feigl, H. and M. Scriven (eds.), *Minnesota Studies in the Philosophy of Science*, volume 1 (University of Minnesota Press, Minneapolis).

Carterette, E. and Friedman, M. (eds.) (1973): *Handbook of Perception*, volume 3, *Biology of Perceptual Systems*, (Academic Press, New York).

________________________________ (1975): *Handbook of Perception*, volume 5, *Seeing*, (Academic Press, New York).

_____ (1978): *Handbook of Perception*, volume 8, *Perceptual Coding*, (Academic Press, New York).
Cartwright, N. (1980): "Measuring Position Probabilities," in Suppes, P. (ed.), *Studies in the Foundations of Quantum Mechanics*, (Philosophy of Science Association, East Lansing).
_____ (1983): *How the Laws of Physics Lie*, (Clarendon, Oxford).
Causey, R. (1979): "Theory and Observation," in Asquith, P. and Kyburg, H. (eds.), *Current Research in Philosophy of Science*, (Philosophy of Science Association, East Lansing).
Churchland, P. (1979): *Scientific Realism and the Plasticity of Mind*, (Cambridge University Press, Cambridge).
Churchland, P. and Hooker, C. (eds.) (1985): *Images of Science*, (University of Chicago Press, Chicago).
Drell, S. (1978): "When is a Particle?" *Physics Today*, June 1978.
Dretske, F. (1969): *Seeing and Knowing*, (University of Chicago Press, Chicago).
_____ (1981): *Knowledge and the Flow of Information*, (MIT Press, Cambridge).
Duhem, P. (1906): *The Aim and Structure of Physical Theory*, Wiener, P. (translator), (Atheneum, New York).
Eisenbub, L. (1971): *The Conceptual Foundations of Quantum Mechanics*, (Van Nostrand, New York).
Feynman, R., Leighton, R., Sands, M. (1965): *The Feynman Lectures on Physics*, volume 3, (Addison-Wesley, Reading).
Fine, A. (1973): "Two Problems of Quantum Measurement," in Suppes, P., Henkin, L., Joja, A., Moisil, G., (eds.), *Logic, Methodology and Philosophy of Science IV*, (North-Holland, Amsterdam).
_____ (1984): "And Not Anti-Realism Either," *Nous* **18**, 51-65.
Fodor, J. (1984): "Observation Reconsidered," *Philosophy of Science* **51**, 23-43.
Ford, K. (1963): *The World of Elementary Particles*, (Blaisdell, New York).
Foss, J. (1984); "On Accepting van Fraassen's Image of Science," *Philosophy of Science* **51**, 79-92.
Frauenfelder, H. and Henley, E. (1974): *Subatomic Physics*, (Prentice-Hall, Englewood Cliffs).
Fritzsch, H. (1983): *Quarks*, (Basic Books, New York).
Galison, P. (1985): "Bubble Chambers and the Experimental Workplace," in Achinstein and Hannaway (1985), 309-373.
Georgi, H. and Glashow, S. (1980): "Unified Theory of Elementary-Particle Forces," *Physics Today*, September 1980, 30-39.

Glymour, C. (1977): "Indistinguishable Space-Times and the Fundamental Group," In Earman, J. Glymour, C. and Statchel, J. (eds.), *Minnesota Studies in the Philosophy of Science*, volume 8, (University of Minnesota Press, Minneapolis).

Grandy, R. (ed.) (1973): *Theories and Observation in Science*, (Prentice-Hall, Englewood Cliffs).

Hacking, I. (1983): *Representing and Intervening*, (Cambridge University Press, Cambridge).

Hanson, N.R. (1958): *Patterns of Discovery*, (Cambridge University Press, Cambridge).

_________ (1969): *Perception and Discovery*, (Freeman, San Francisco).

Hawking, S. (1976): "Breakdown of predictability in gravitational collapse," *Physical Review D* **14**, #10, 2460-2473.

Hawking, S. and Ellis, G. (1973): *The Large Scale Structure of Space-Time*, (Cambridge University Press, Cambridge).

Heisenberg, W. (1949): *The Physical Principles of Quantum Theory*, C. Eckart and F. Hoyt (trans.), (Dover, New York).

Hensel, H. (1973): "Temperature Recption," in Carterette and Friedmen (1973), 317-325.

Howard, I. (1973): "The Spatial Senses," in Carterette and Friedman (1973), 233-290.

Hren, J., Goldstein, J., and Joy, D., (eds.) (1979): *Introduction to Analytic Electron Microscopy*, (Plenum Press, New York).

Isaacson, M., Ohtsuki, M., and Utlaut, M. (1979): "Electron Microscopy of Individual Atoms," in Hren, et. al., (1979). 343-368.

Jammer, M. (1974): *The Philosophy of Quantum Mechanics*, (Wiley, New York).

Jauch, J. (1968): *Foundations of Quantum Mechanics*, (Addison Wesley, Reading).

Kemble, E. (1973): *The Fundamental Principles of Quantum Mechanics*, (Dover, New York).

Kuhn, T. (1970): *The Structure of Scientific Revolutions*, second edition, (University of Chicago Press, Chicago).

Laudan, L. (1981a): "A Confutation of Convergent Realism," *Philosophy of Science* **48**, 19-49.

_________ (1981b): "The Medium and its Message: A Study of Some Philosophical Controversies about Ether," in Cantor and Hodge (1981), 157-185.

Lehninger, A. (1970): *Biochemistry*, (Worth, New York).

Lewis, D. (1970): "How to Define Theoretical Terms," *Journal of Philosophy* **67**, 427-446.

Lodge, O. (1925): *Ether and Reality*, (George Doran Co., New York).

Malament, D. (1977): "Observationally Indistinguishable Space-Times," in Earman, J., Glymour, C., and Statchel, J. (eds.), *Minnesota Studies in the Philosophy of Science*, volume 8, (University of Minnesota Press, Minneapolis).

Margenau, H. (1950): *The Nature of Physical Reality*, (McGraw-Hill, New York).

__________ (1958): "Philosophical Problems Concerning the Meaning of Measurement in Physics," *Philosophy of Science* **25**, 23-33.

Maxwell, G. (1962): "The Ontological Status of Theoretical Entities," in Feigl, H. and Maxwell, G. (eds.), *Minnesota Studies in the Philosophy of Science*, volume 3, (University of Minnesota Press, Minneapolis).

Nagel, E. (1961): *The Structure of Science*, (Harcourt, Brace and World, New York).

Narlikar, J. (1983): *Introduction to Cosmology*, (Jones and Bartlett, Boston).

Penrose, R. (1969): "Gravitational Collapse: The Role of General Relativity," *Rivisita Nuovo Cimento*, Ser. 1, 252-276.

Peyrou, C. (1967): "Bubble Chamber Principles," in Shutt (1967), 19-58.

Putnam, H. (1962): "What Theories Are Not," in Nagel, E., Suppes, P. and Tarski, A. (eds.), *Logic, Methodology and Philosophy of Science*, (Stanford University Press, Stanford).

Reichenbach, H. (1944): *Philosophical Foundations of Quantum Mechanics*, (University of California Press, Berkeley).

Rosenbaum, D. (1975): "Perception and Extrapolation in Velocity and Acceleration," *Journal of Experimental Psychology: Human Perception and Performance* **1**, 395-403.

Rynasiewicz, R. (1984): "Observability," in Asquith, P. and Kitcher P. (eds.), *PSA 1984*, volume 1.

Schlegel, R. (1980): *Superposition and Interaction*, (University of Chicago Press, Chicago).

Sekuler, R. (1975): "Visual Motion Percpetion," in Carterette and Friedman (1975), 387-430.

Shapere, D. (1982): "The Concept of Observation in Science and Philosophy," *Philosophy of Science* **49**, 485-525.

Shrader-Frechette, K. (1982): "Quark Quantum Numbers and the Problem of Microphysical Observation," *Synthese* **50**, 125-146.

Shutt, R. (ed.) (1967): *Bubble and Spark Chambers*, (Academic Press, New York).

Tipler, F. (1985): "Note on Cosmic Censorship," *General Relativity and Gravitation* **17**, #5, 499-507.

van Fraassen, B. (1980): *The Scientific Image*, (Oxford University Press, Oxford).

Weinberg, S. (1972): *Gravitation and Cosmology*, (Wiley, New York).

Winnie, J. (1967): "The Implicit Definition of Theoretical Terms," *British Journal for the Philosophy of Science* **18**, 223-229.

Wischnitzer, S. (1970): *Introduction to Electron Microscopy*, second edition, (Pergamon, New York).

Wyburn, G. (ed,) (1964): *Human Senses and Perception*, (Oliver and Boyd, Edinburgh).

INDEX